THIRD EDITION

THE
MOON
BOOK

FASCINATING FACTS ABOUT THE
MAGNIFICENT, MYSTERIOUS MOON

KIM LONG

SCIENCE ADVISOR

LARRY SESSIONS

FORMER DIRECTOR, NOBLE PLANETARIUM (FORT WORTH, TEXAS)
FORMER STAFF ASTRONOMER, GATES PLANETARIUM (DENVER, COLORADO)

Original cover design: Debra B. Topping

Cover photography: Lick Observatory

All illustrations by the author unless otherwise noted.

ISBN 978-0-9911266-4-4

CONTENTS

INTRODUCTION

The birth of this book dates to 1988, when it was conceived as a companion volume to the author's *Moon Calendar*, which was created in 1981 and continues as an annual. The current volume marks the third edition of *The Moon Book*, building and expanding on the work of two previous editions. In the time span of almost two decades, not much has changed on the Moon, but significant developments have affected the worlds of personal computing, publishing, and astronomy, all factors that have an impact on the contents of this new edition.

These days, information is a click away on the Internet, with a "click" as often originating from a portable device as a desktop computer and apps provide instant, localized details about the changing cycles of nature, including the Moon. Ebooks have wrought considerable upheaval in the traditional world of printed books, some beneficial and some that have yet to be fully manifested. In addition, India, Japan, and China have entered the space age with rockets and spacecraft that add energy and enthusiasm for astronomy.

Through all of this, however, people remain curious about the patterns of nature, including the ages-old systems of the stars and planets over our heads. This curiosity has spanned millennia, with relics such as Stonehenge still exciting interest and passion. Within the pages of this book, written and graphic descriptions provide support for the questions that linger about the celestial satellite that is over our heads, from the most basic concepts about the structure of the Moon to the recognizable and orderly patterns of its movement.

Advanced probes, scientific analysis, computerized number crunching, and even first-person observations on the surface—all on the upswing in recent years—have not diminished our curiosity about the Moon, but stirred additional passion and involvement. With the increasing probability that a return of human explorers to the lunar surface will happen, for commercial development and as a base for further exploration of the solar system, people have more questions than ever.

The Moon Book—or any single book about the Moon—cannot include answers to everything that is asked about the Moon. Our endeavor in 1988 provided a starting point and with this third edition, we hope it continues to provide readers with useful support for their lunar curiosity.

THE MOON'S ORBIT

The Moon orbits around the Earth in an elliptical path. This path is not a circle, but an elongated circle. The degree to which an elliptical path is elongated is called its eccentricity. The eccentricity of the Moon's orbit is small, too small to accurately depict in an illustration that would fit in this book, but larger than the elliptical orbit of the Earth around the Sun, or that of the other planets. The difference between the shortest (perigee) and longest (apogee) distances from the Earth to the Moon within its orbit is only about the width of four Earths (see page 135); it varies from about 7% less than the mean distance at its closest to about 6 percent more than the mean at its farthest. In technical terms, the mean eccentricity of the Moon's orbit is 0.0549.

Over time, the elliptical shape of the Moon's orbit shifts slightly, causing the shortest and farthest distance from the Earth in any given year to change. In the 20th century, the greatest distance was 252,731 miles (406,712 km) on March 2, 1984, and the shortest distance was 221,451 miles (356,375 km) on January 4, 1912. In the 21st century, the greatest distance will be 252,160 miles (406,709 km) on December 12, 2061. The shortest distance will be 220,984

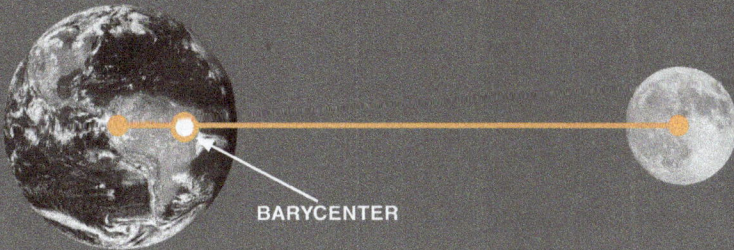

EARTH-MOON BARYCENTER

The combined masses of two bodies in a binary system have a common center of mass, called the barycenter. The barycenter of the Earth-Moon system is located about 3,000 miles out from the center of the Earth (about 900 miles under the surface). This location changes constantly because the distance between the Earth and the Moon is also changing constantly as the Moon orbits in an elliptical path around the Earth.

BARYCENTER

miles (356,425 km) on December 6, 2052. Most of the time and for most purposes, the distances from Earth in the orbit are averaged, with a mean perigee and apogee used to express these extremes.

The orbit of the Moon around the Earth, is more complex than just that of one body moving around another. The Moon and the Earth, which each has its own center of gravity, orbit around a common center of gravity called a barycenter. A barycenter is generated by a system formed of two masses. The

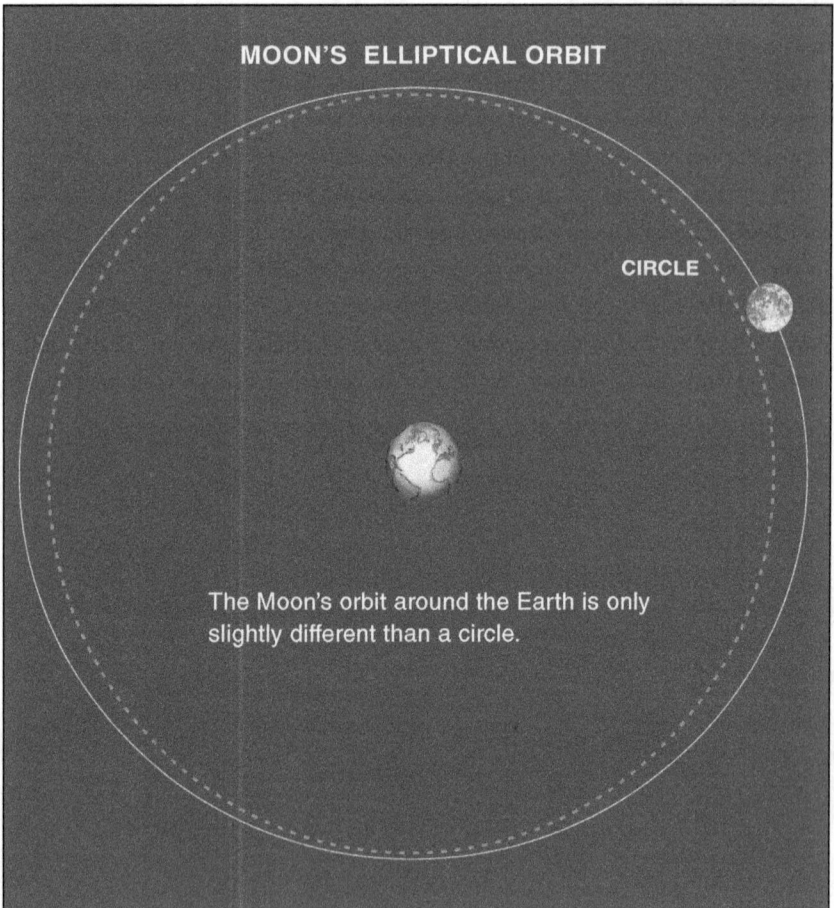

MOON'S ELLIPTICAL ORBIT

CIRCLE

The Moon's orbit around the Earth is only slightly different than a circle.

THE ROTATION OF THE MOON

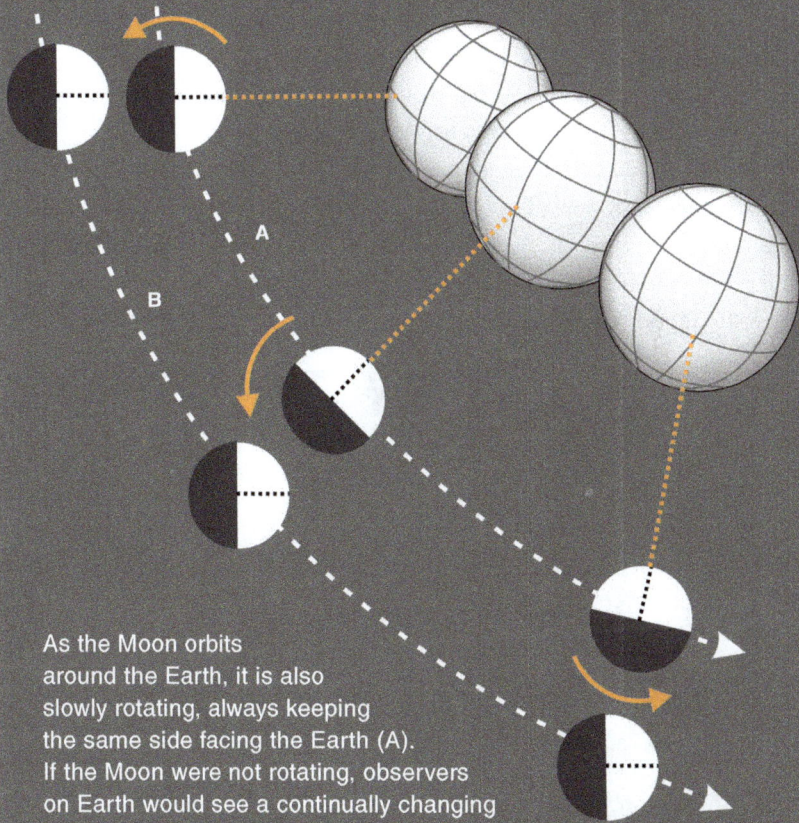

As the Moon orbits
around the Earth, it is also
slowly rotating, always keeping
the same side facing the Earth (A).
If the Moon were not rotating, observers
on Earth would see a continually changing
surface (B). If the Moon were not rotating,
during every orbit around the Earth, the far side
would come into view, something that does not happen.

barycenter of the Earth-Moon system is in constant motion because both bodies are themselves in motion, one orbiting around the other, and both rotating around their axis.

The Earth, being much larger than the Moon, "pulls" the barycenter much closer to itself than to the Moon, well within its own body. The barycenter is located below the surface of the Earth, about three quarters out (3,000 miles) from the center of the planet, but this position shifts as the two bodies move through space and the distance between them changes.

When the Moon is farthest away in its orbit, the barycenter is also at its extreme distance, only 869 miles (1400 km) below the Earth's surface. When the Moon is closest in its orbit, the barycenter moves closer the center of the Earth, about 1,180 miles (1900 km) below the surface. Because the Moon's orbit is an ellipse, the motion of the barycenter over time also forms an ellipse.

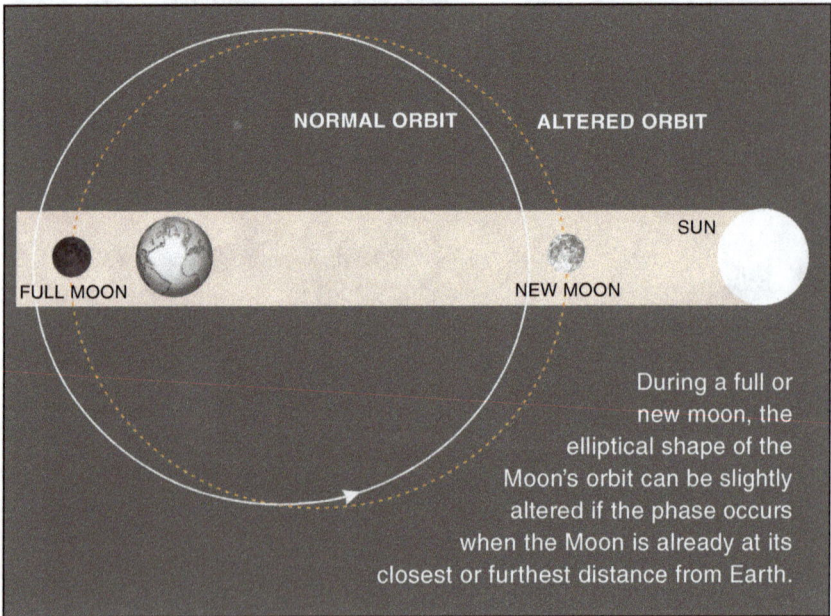

NORMAL ORBIT   ALTERED ORBIT

SUN

FULL MOON   NEW MOON

During a full or new moon, the elliptical shape of the Moon's orbit can be slightly altered if the phase occurs when the Moon is already at its closest or furthest distance from Earth.

MOON SPEED

Relative to the Earth, the Moon makes one rotation around its axis every 29½ days, on average. This is the same time it takes for the Moon to complete one revolution around the Earth. The similarity of these two figures is no coincidence. The Moon rotated much faster in the past, but one of the effects of the Earth's gravity on the Moon over millions of years has been to slow down the rotation until the Moon has become "locked in step" with the Earth (see page 3). The technical term is synchronous rotation.

The Earth rotates at about 1,000 miles an hour as measured from a point on the surface at the equator, while the Moon rotates at only 10 miles an hour. By comparison, the orbital speed of the Moon is much faster than its rotation. The average speed of the Moon on its monthly trip around the Earth is 2,287 miles an hour (3,683 kilometers/hour). However, the Moon's orbit is not a circle but an ellipse. The effect of an elliptical path on the speed of an orbiting object is

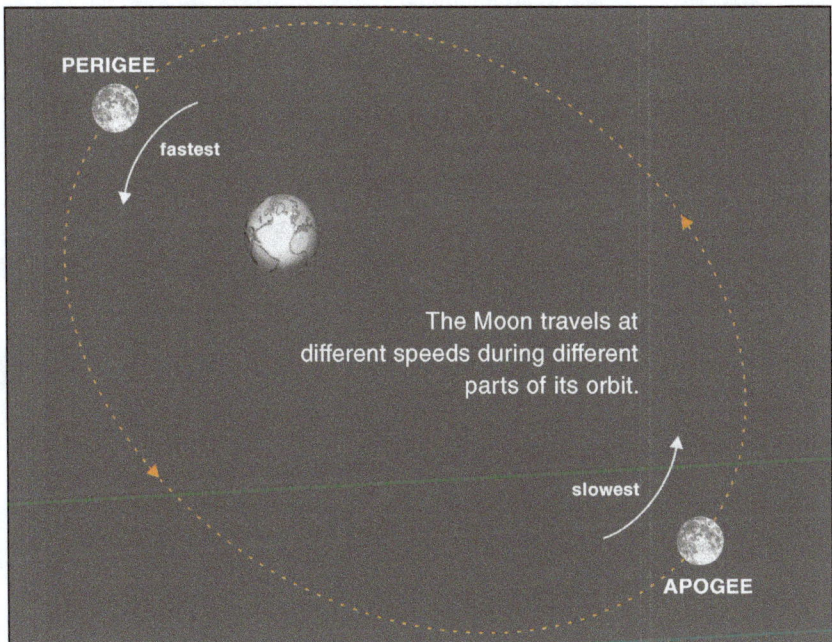

PERIGEE

fastest

The Moon travels at different speeds during different parts of its orbit.

slowest

APOGEE

to change the speed at different segments of the orbit. When the Moon is closest to the Earth (its perigee), it is traveling at its maximum speed, 2,402 miles an hour (3874 km/hr). At the farthest point from the Earth (the apogee) the speed is slowest, 2,151 miles an hour (3470 km/hr).

An observer on the surface of the Earth sees this orbital speed as the movement of the Moon across the sky at about one full Moon's width per hour; the average distance over a 24-hour day is a little over 13 degrees (13°11', to be exact). At its theoretical fastest, the Moon could cover as much as 15°24' per day, an action that would require a combination of circumstances, primarily an extremely close perigee that is very close to the time of the full or new moon phase. At its slowest extreme speed, the Moon is less influenced by outside forces. During most lunar months, its slowest progress is just under 12 degrees of motion a day.

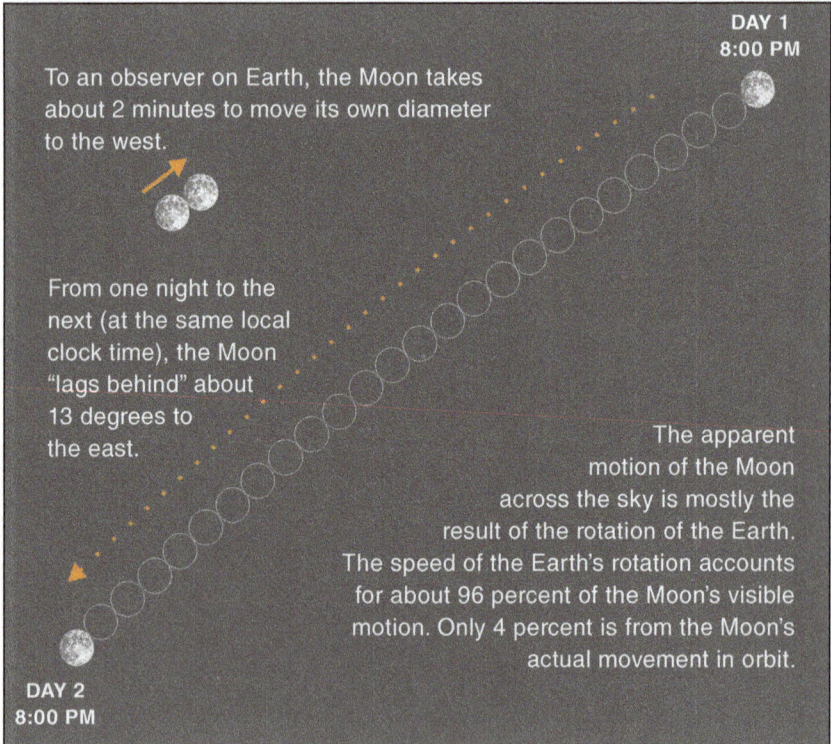

DAY 1
8:00 PM

To an observer on Earth, the Moon takes about 2 minutes to move its own diameter to the west.

From one night to the next (at the same local clock time), the Moon "lags behind" about 13 degrees to the east.

The apparent motion of the Moon across the sky is mostly the result of the rotation of the Earth. The speed of the Earth's rotation accounts for about 96 percent of the Moon's visible motion. Only 4 percent is from the Moon's actual movement in orbit.

DAY 2
8:00 PM

THE LUNAR MONTH

The Moon completes one orbit around the Earth in about 29½ days. This period is called a lunation, lunar month, or synodic month. A lunation begins at the exact time of a new moon and ends at the exact time of the next new moon. One lunation is officially 29 days, 12 hours, 44 minutes, 2.8 seconds long. However, this measurement is an average, not a constant, and reflects monthly variations that occur over a long period of time (see page 20).

If the lunar cycle is measured by timing the orbit in relationship to the position of a specific star, it is called a sidereal cycle. The sidereal lunar month is only 27 days, 7 hours, 43 minutes, 11.5 seconds long. The difference between the synodic month and the sidereal month is about 2 days per calendar month, which is about how much a full moon will "lag" behind the calendar, although the variation in length of a calendar month makes this an uneven rule of thumb.

A lunation is a very visible cycle that is easy to observe. Most people, however, notice the full moon instead of the new moon because the former is visible and the latter is not; an unofficial lunar cycle could be considered one full moon to the next full moon. The exception is for traditional lunar calendars, most of which begin their months at the time of the new moon. Since the new moon is not visible, tradition dictates that the sighting of the first crescent moon marks the beginning of the month. For more information on lunar calendars, see page 104.

Lunations are numbered in sequence. The sequence began with Lunation Number 1, designated by astronomers around the world as beginning with the new moon on January 16, 1923. There are 13 lunations in every calendar year because calendar months are longer than lunar months, with the exception of February.

In this century, the longest lunation is 29 days, 19 hours, 47 minutes and begins with the new moon on December 18, 2017. The shortest will be 29 days, 6 hours, and 35 minutes and begins on June 16, 2053.

THE ECLIPTIC

The Sun's path across the sky—actually caused by the revolution of the Earth around the Sun—is called the ecliptic. The ecliptic is a kind of road map relative to the stars, creating a guide to the sky that can be used both to find directions and tell time. Prominent stars and constellations are signposts on this path, which also includes the twelve traditional constellations marking the zodiac, used by both astronomers and astrologers, at least those in western cultures. Even though the stars are the same, other traditional and religious astrologies describe different constellations and zodiac groupings along the ecliptic (see page 8).

The Moon's path is tilted to the ecliptic by about 5.15 degrees and since this tilt is itself rotating, the lunar path over time will eventually "sweep out" an area that is roughly 5 degrees above and below the path of the Sun. The point where the crosses the ecliptic from south to north is called the ascending node; the descending node marks the spot on the ecliptic where it crosses north to south. The nodes are important, because it is only when the Moon crosses a node that an eclipse can occur.

Starting at its ascending node, it takes the Moon an average (mean) of 27 days, 5 hours, 5 minutes, and 36 seconds to get back where it started, a period called the draconic month. This draconic month can vary from its mean by more than six hours.

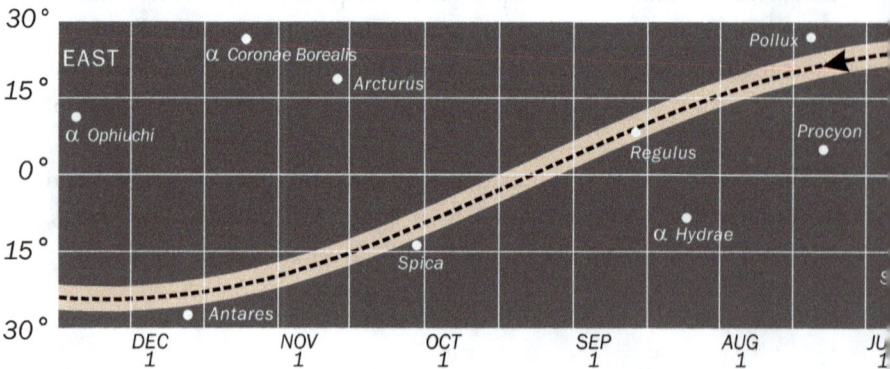

8

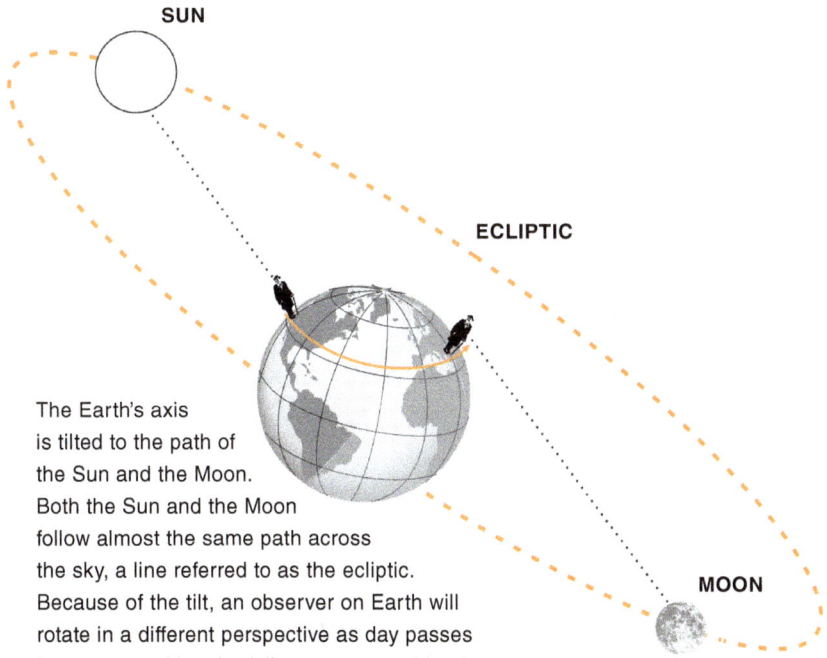

SUN

ECLIPTIC

MOON

The Earth's axis
is tilted to the path of
the Sun and the Moon.
Both the Sun and the Moon
follow almost the same path across
the sky, a line referred to as the ecliptic.
Because of the tilt, an observer on Earth will
rotate in a different perspective as day passes
into night, making the full moon appear high in
the sky at midnight when the Sun was low at noon,
and vice versa.

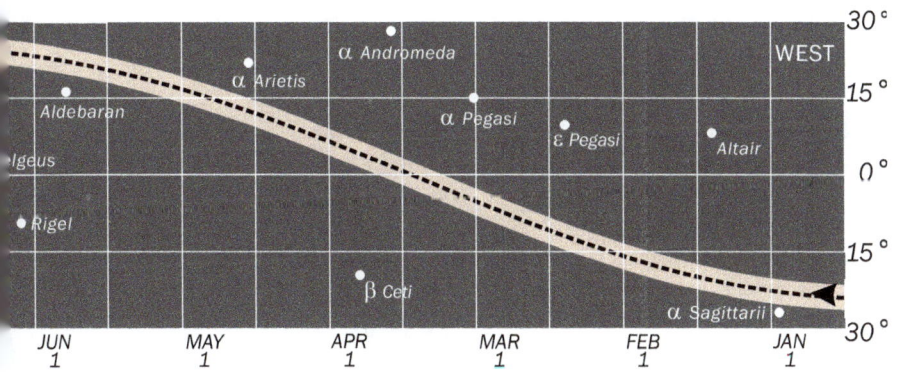

9

The plane of the Moon's orbit is tilted away from the plane of the Earth's orbit by about 5 degrees. The two points where the planes intersect are called node. One is an ascending node and the other is a descending node.

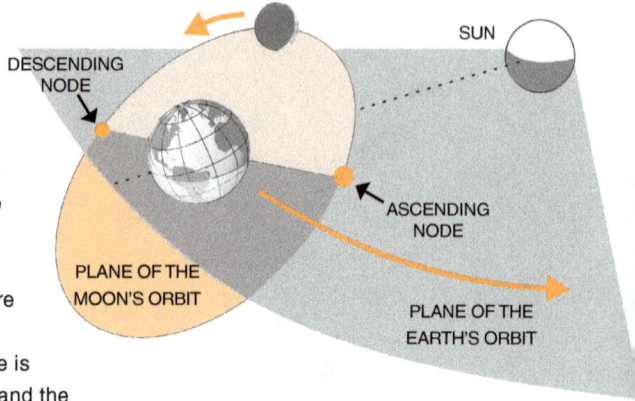

DESCENDING NODE

SUN

ASCENDING NODE

PLANE OF THE MOON'S ORBIT

PLANE OF THE EARTH'S ORBIT

The plane of the Moon's orbit rotates slowly around the Earth. If the nodes happen to fall on the line directly between the Earth and the Sun, a full moon or a new moon will result in an eclipse.

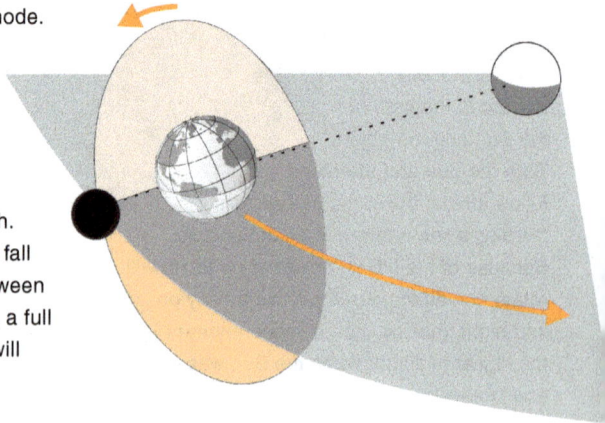

When the new moon is at the node between the Sun and the Earth, the result is a solar eclipse. A lunar eclipse results when a full moon is at the opposite node.

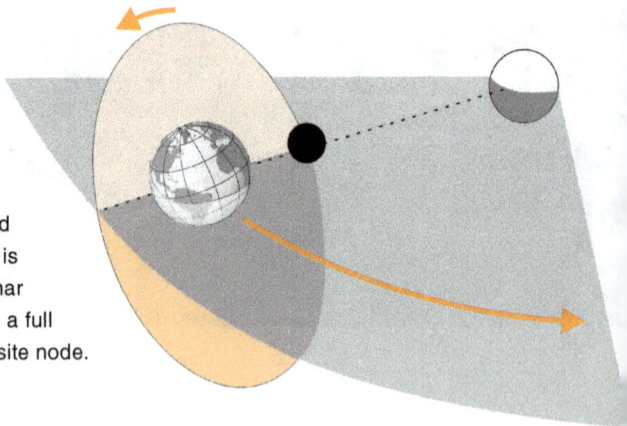

LUNAR PHASES

The Moon appears to change shape as it moves through its monthly cycle. The Moon itself is not changing shape, however, only the part that is illuminated by the Sun, giving us the familiar and distinctive lunar phases. The four official phases that are included in calendars and almanacs are, in order, new moon, first quarter moon, full moon, and last quarter moon.

The new moon is actually invisible, as it is defined as an instant of time when the Moon is between the Sun and the Earth and therefore lost in the bright light of the Sun. At this stage, not even a thin crescent is illuminated by the Sun and when the Moon is approaching and leaving the new moon phase, there is a period of two to four days when it cannot be easily seen for the same reason—too much light from the Sun—and a period of at least a day when it can't be seen at all.

The first and last quarter moons mark the halfway points between the new moon and full moon. The first quarter moon is always "first" and is distinguished by the illuminated half of the lunar surface being on the right-hand side. The last quarter moon is the reverse, with the illuminated half on the left-hand side.

The new moon, first quarter moon, full moon, and last quarter moon (also known as the third quarter moon) are called the primary phases and each represents a single point in time, though we often refer to each phase by the calendar day it fall on. The periods in between the primary phases—waxing crescent moon, waxing gibbous moon, waning gibbous moon , and waning crescent moon—are called the intermediate phases and span about a week.

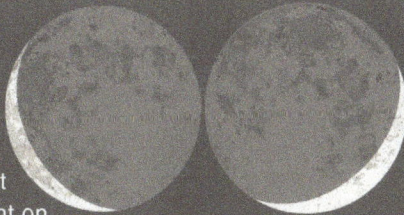

The old crescent moon shows light on the left and is seen in the east in early morning before the Sun sets.

The new crescent moon shows light on the right and is seen in the west in early evening after the Sun sets.

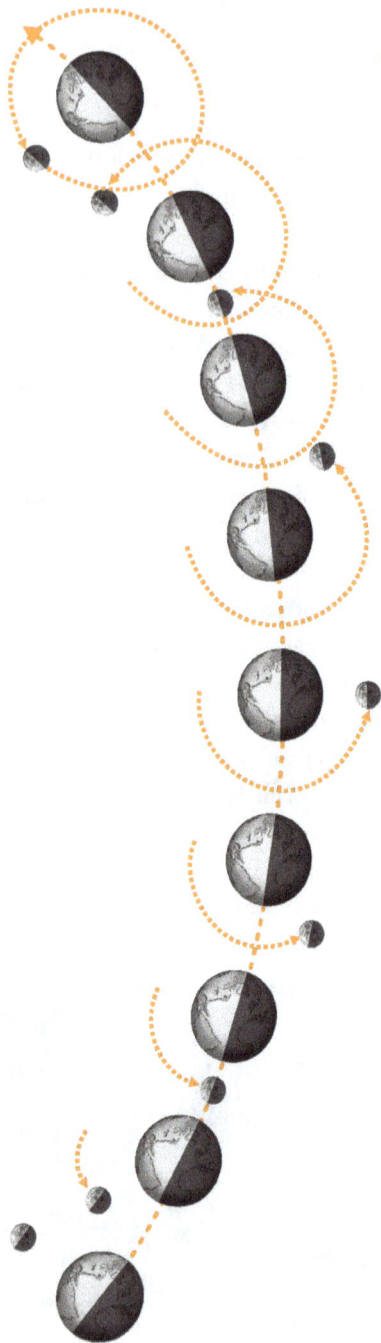

New Moon
One lunar cycle has been completed.

Waning Crescent Moon
The Moon is ahead of the Sun but beginning to move toward the trailing edge of the orbit.

Last Quarter Moon
The Moon is ¾ around its orbit, on the leading edge of the Earth's orbit.

Waning Gibbous Moon
The Moon is moving in to the leading side of the Earth's orbit.

Full Moon
The Moon is opposite the Sun, on the far side of the Earth.

Waxing Gibbous Moon
The Moon is moving into the far side of the Earth's orbit, away from the Sun.

First Quarter Moon
The Moon is ¼ around the Earth, still on the trailing side.

Waxing Crescent Moon
The Moon is trailing behind the orbit of the Earth.

New Moon
The Moon is between the Sun and the Earth.

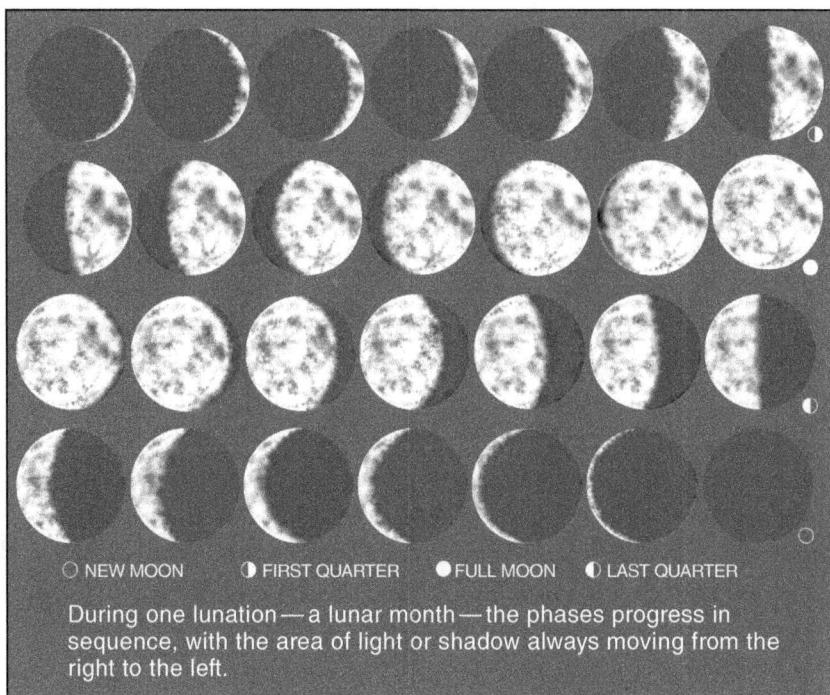

During one lunation — a lunar month — the phases progress in sequence, with the area of light or shadow always moving from the right to the left.

The lighted part of the Moon always points the way to the Sun. If the lighted half is on the right, the Sun is on the right (west), meaning the Sun is ahead of the Moon. If the left half is lighted, the Moon is ahead of the Sun, and the Sun is on the left (east).

The sequence of the lunar phases always proceeds with the lighted part of the Moon growing from right to left until the full moon, then receding from right to left until the new moon (in the Northern Hemisphere). If the Moon is light on the right side, the light will be expanding to the left; if the Moon is dark on the right side, the shadow will be expanding to the left.

The new moon is also called the "dark of the moon," as that is when it is totally dark. The period of darkness officially lasts for only a fraction of a second, marking the point when the Moon is directly between the Sun and the Earth. However, observers on Earth don't see any "slivers" of light at the edges of the Moon just before and after this point because the new moon is too close to the

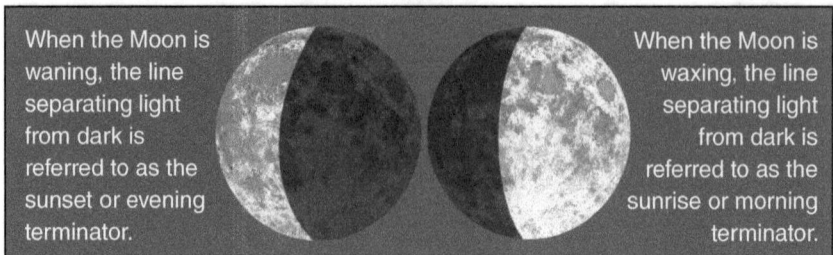

When the Moon is waning, the line separating light from dark is referred to as the sunset or evening terminator.

When the Moon is waxing, the line separating light from dark is referred to as the sunrise or morning terminator.

Sun to see anything clearly. Except during eclipses, when the body of the Moon comes directly between the Earth and the Sun, or vice versa, the New Moon will be within 5 degrees above or below the Sun.

During the period around the new moon, the Moon follows the Sun very closely; immediately after the new moon, it begins to lag behind. The first visible crescent moon (with the crescent on the right) is usually spotted two or three days after the new moon. Observers can see this young crescent moon, as it is sometimes called, just after sunset, with the Moon following the Sun down over the western horizon; there has to be at least 7 degrees of separation between the two bodies for the crescent to be visible, and usually more. The earliest that a "naked eye" observer on Earth has ever seen the young crescent moon is about 14 hours after new moon (see page 52).

As the Moon "grows," or waxes, the lighted portion on the right gradually increases until it forms an almost perfect half circle, the first quarter moon. Even though half of the surface is illuminated, it is called a quarter moon because of the unseen far side—only 25 percent of the total sphere is illuminated.

Quarter moons are created by the position of the Moon in relation to the Earth and Sun. At this point in the cycle, the orbit of the Moon has moved it to a position off to the side of the line connecting the Earth and the Sun. At the point of first quarter moon, the Moon is 90 degrees (perpendicular) to this imaginary line. Technically, this phase is called a quadrature.

The edge of light that gradually creeps across the lunar surface is called the terminator. The terminator is rarely an even line, as the rough surface and uneven curvature of the Moon distort it, a phenomenon visible through binoculars or telescopes and even at times to the unaided eye. At the point of first and last quarter phase, naked eye observations from Earth usually show a reasonably

COMPARING MOON PHASES

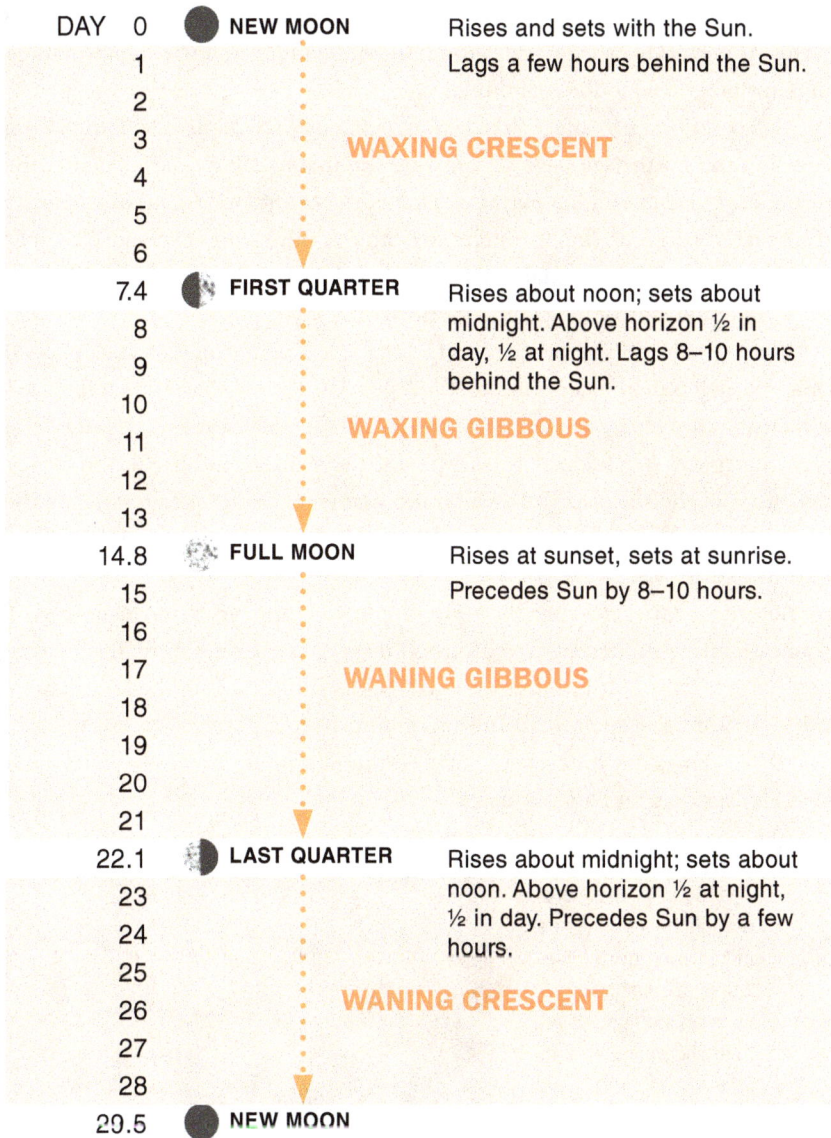

DAY	0	NEW MOON	Rises and sets with the Sun.
	1		Lags a few hours behind the Sun.
	2		
	3		**WAXING CRESCENT**
	4		
	5		
	6		

	7.4	FIRST QUARTER	Rises about noon; sets about midnight. Above horizon ½ in day, ½ at night. Lags 8–10 hours behind the Sun.
	8		
	9		
	10		
	11		**WAXING GIBBOUS**
	12		
	13		

	14.8	FULL MOON	Rises at sunset, sets at sunrise. Precedes Sun by 8–10 hours.
	15		
	16		
	17		**WANING GIBBOUS**
	18		
	19		
	20		
	21		

	22.1	LAST QUARTER	Rises about midnight; sets about noon. Above horizon ½ at night, ½ in day. Precedes Sun by a few hours.
	23		
	24		
	25		
	26		**WANING CRESCENT**
	27		
	28		
	29.5	NEW MOON	

straight perpendicular line, but the non-smooth surface keeps this from being absolutely straight. In most modern depictions of the lunar cycle, a give-away to their being generated by computer graphics programs is an absolutely straight terminator used for the quarter moons—photographs and traditional renderings are more true to the real thing.

Because librations affect exactly what portion of the front face of the Moon is facing the Earth (see page 63), during the first and last quarter phase, the terminator's exact location may vary. That is, the central meridian, which marks the exact north-south line through the center of the Moon's front surface, may not line up with the terminator. At extremes, the terminator may appear up to 7°45' on either side of the central meridian.

As the shadow line moves across the surface, it is moving at a steady speed of about half a degree an hour, or 12.2° per day. If the Moon were a flat, two-dimensional object, this shadow would appear to move across the surface at a regular pace. But because it is a sphere, the shadow line is affected by the curvature of the surface. To Earth-bound observers, it moves fastest across the surface between the first quarter and last quarter moon, when more of the curvature is visible.

Before and after this time, however, the shadow appears to move more slowly because the rounded surface drops off more quickly. Therefore, from day to day, observers on Earth will see more day-to-day differences in the terminator's position from first quarter to full moon and from full moon to last quarter, and fewer day-to-day differences between the new moon and first quarter and between last quarter and new moon.

The Moon is visible during daylight hours during part of the lunar month. The first quarter moon, as seen on the right, rises about noon and sets about midnight.

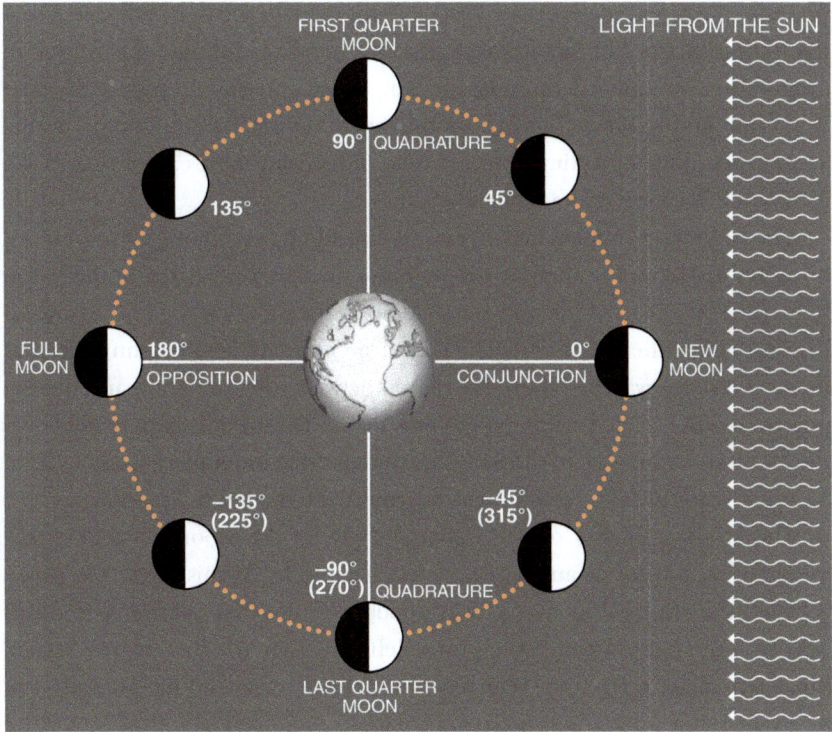

FIRST QUARTER
MOON

LIGHT FROM THE SUN

90° | QUADRATURE

135°

45°

FULL
MOON

180°

OPPOSITION

0°

CONJUNCTION

NEW
MOON

−135°
(225°)

−45°
(315°)

−90°
(270°) | QUADRATURE

LAST QUARTER
MOON

The traditional form of diagramming the Moon's phases.

As the Moon continues waxing, it falls farther and farther behind the movement of the Sun. In the period after the first quarter moon the terminator continues to move to the right, lighting more and more of the surface. This period is referred to as the waxing moon; it is also traditionally called the gibbous moon (which can also refer to the period after the full moon up until the last quarter moon), or more specifically, the waxing gibbous moon. The visible feature of a gibbous moon is a "bulge" in the shape of the lighted portion. This period lasts from the first quarter moon until the time of the full moon and then again after the full moon, lasting until the last quarter moon.

The Moon becomes full when the orbital path of the Moon carries it directly opposite of the Sun, which at this point is on the other side of the Earth. In this position, the Moon receives the direct light of the Sun across its full face, forming the distinctive full moon image. Another term for full moon is "moon in opposition," because it is opposite the Sun. Being opposite the Sun at the time of full moon, the Moon rises just as the Sun is setting. The full moon occurs at an exact moment in time, but the time of its rising varies according to the geographic location on an observer.

The full moon immediately begins diminishing, even though this may not be visibly apparent for a day or two, because the spherical shape of the Moon hides the first effects of the growing shadow on the right-hand side (the same is true just before the full moon). The period after the full moon up until the new moon is called a waning moon; the period between the full moon and the last quarter moon is called a waning gibbous moon. During this part of the cycle, the Moon is catching up to the Sun. The sunlight that illuminates it comes from "behind"—to the left—so the shadow grows from the left to the right.

The growing shadow on the right-hand side of the Moon's face gradually increases until it forms an almost perfect half circle. At this point, the phase is called the last quarter moon. It is opposite to the first quarter moon, this time with the lighted half being on the left side. Like the first quarter moon, this phase is also technically referred to as a quadrature, and the Moon is now positioned at a right angle (perpendicular) to the imaginary line connecting the Sun and the Earth.

The shadow on the Moon continues to diminish after the last quarter, with the growing shadow obscuring more and more of its face; this period is called the waning crescent moon. In less than a week, the entire face will be in shadow, and the cycle will be back at the beginning, the new moon. The last part of the visible lunar phase, between the last quarter moon and the new moon is often called the old crescent moon.

Beginning with the new moon, the passage of time in a lunar cycle is often marked with a number called the Moon's "age." This figure is usually represented with a whole number (for the number of calendar days) or a decimal number—6.5 or 16.3, for example—to represent the time elapsed.

These NASA photographs provide a detailed look at the position and pattern of the terminator, the line that marks the separation of the illuminated region from the region in shadow. Traditionally, graphic depictions of the quarter moons depict this phase with a sharp line running vertically across the Moon's surface.

Computer graphic programs that model realistic conditions, using formulas to apply shadows to a three-dimensional artificial moon, usually repeat this convention, but can provide attractive and reasonably accurate results. Moon-phase clocks and watches, on the other hand, provide the least accurate images of phases, as they are limited to rotating a fixed-diameter, two-dimensional circle—representing the shadowed portion—in front of another circle, as depticted here. Except for a full moon, this system never portrays what a moon phase actually looks like.

As the photographs on the right show, the terminator is less of a sharp line and more of an irregular vertical band, indicating the rough surface where the light falls. Craters, mountains, and other features of the Moon's surface also interfere with what would otherwise be a simple, clean line running across the surface of a sphere.

The terminator at various parts of the lunar cycle displays different characteristics. TOP RIGHT First quarter phase. CENTER RIGHT Waxing crescent. BOTTOM RIGHT Waxing gibbous. Photographs courtesy NASA.

OUT OF PHASE

One lunar cycle, from new moon to new moon, is completed about every 29½ days. The "official" length of this period is 29.53059 days, or 29 days, 12 hours, 44 minutes, and 2.8 seconds. Many people think of this figure as a constant, a regular benchmark that marks out lunar rhythm over time. Unfortunately, like many things associated with the Moon's movement, the truth involves a lot of variation.

During its monthly orbit around Earth, the Moon's speed is affected by many factors, including the gravitational forces of the Earth and the Sun and regular and irregular variations in its own orbit. The major effect on the length of a lunation comes from the gravitational force of the Sun.

From one month to the next, a lunation—the technical name for the period from one new moon to the next new moon—can vary from a few minutes to several hours, with an average variation of about fourteen hours over 100 years (13.44 hours, to be exact). Roger Sinnott, an editor at *Sky & Telescope* magazine, calculated the variation over a much longer period of 500 years (from 1600 to 2100) and found that the mean variation was about nine hours. From 1900 to 2100, the extremes range from about six hours less than the average to a

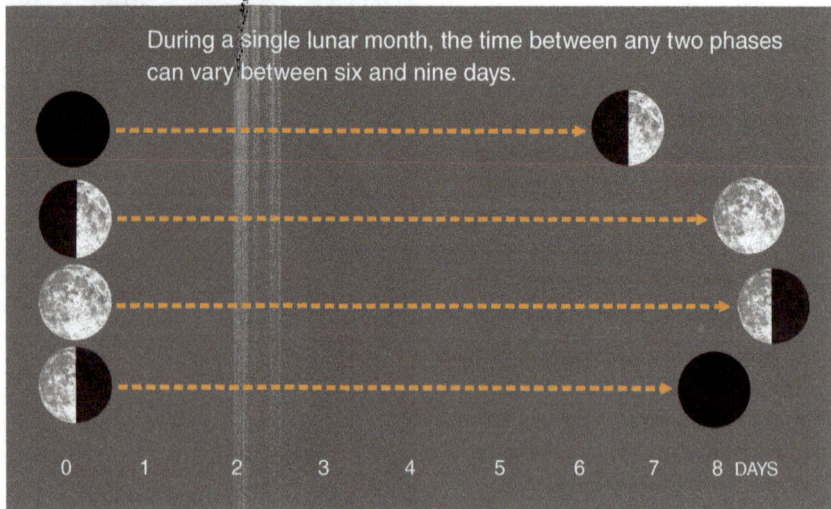

During a single lunar month, the time between any two phases can vary between six and nine days.

0 1 2 3 4 5 6 7 8 DAYS

little more than seven hours greater than the average, according to calculations made by Jean Meeus, a noted Belgian mathematician and expert on astronomical cycles. The shortest lunation occurred on June 25 to July 24, 1903, a period of 29 days, 6 hours, and 35 minutes. The longest lunation was December 24, 1973, to January 23, 1974, a period of 29 days, 19 hours, and 55 minutes. A lunation is at a minimum length when the new moon is closest to the Earth (perigee) and a maximum length when it is farthest from Earth (apogee).

More confounding for moon watchers is the fluctuation in the period of time between any two successive phases, full moon to last quarter moon, for example. This period may vary considerably from one phase to the next because one phase is a quarter of the lunar path, which is not evenly round but an ellipse , and segments of an ellipse vary in length. If these periods were equal, each would span 7.38 days, or one quarter of the synodic month. But instead, they vary by as much as 11 percent from this standard — plus or minus 19 hours — yielding a length of time between two successive phases that can vary between six to nine calendar days, although most periods are not this extreme.

While phases and lunar months fluctuate, particularly compared to civil calendars, there are patterns that appear over longer periods of time. Every 19 years, for example, there is a recurring cycle of 235 lunar months (lunations) when the new moon and full moons fall on the same calendar days (with occasional exceptions). In other words, in 1999, a new moon falls on January 17 and 19 years later in the year 2018, it also falls on January 17, a pattern known as the Metonic Cycle.

The Greek astronomer Meton is credited with first reporting this discovery in 432 BCE. This observation is also sometimes credited to Euctemon, a colleague of Meton, as well scholars in ancient China and Babylonia.

One lunation is about 29.5 days and one year is 365 days. Every 6,939 days or 235 lunations, the two counts coincide, with the same moon phases appearing on the same calendar days. Or almost — because of minor variations in the lunar cycle, there are typically a few days when two years that are Metonic pairs are mismatched, with one or two days not quite aligning. In the paired Moon Calendars shown on page 23, for example, the calendar days for some of the full moons from January to July are offset by a day.

The Metonic cycle was the inspiration for the Golden Number, a system dating to the Middle Ages in Europe, and is still used for determining lunar-based holidays, particularly Easter. Golden Numbers are usually depicted as Roman numerals — I through XIX. Less well known variations on Meton's discovery are the Callipic cycle (covering four Metonic cycles, or 76 years) and the Hipparchic cycle (four Callipic cycles minus a day, or 304 years).

Another pattern was observed and named by early Greek astronomers. Based on an eight-year cycle, this phenomenon is called the *octaeteris*. Every eight years, the new moon and full moon fall on a calendar day that is one to two days later than the day in the first calendar year. In other words, a new moon falls on January 17 in 1999 and eight years later, in 2007, it falls two days later, on January 19. Moon phases also precede each other on almost the same dates every two years. The new moon in 1999, for example, falls on January 17 but two years later there is a last quarter moon one day earlier, on January 16 in 2001.

THE GOLDEN NUMBER

To find a golden number for a year, divide by the year by 19 and add 1 to the remainder.

2015	II	2033	I
2016	III	2034	II
2017	IV	2035	III
2018	V	2036	IV
2019	VI	2037	V
2020	VII	2038	VI
2021	VIII	2039	VII
2022	IX	2040	VIII
2023	X	2041	IX
2024	XI	2042	X
2025	XII	2043	XI
2026	XIII	2044	XII
2027	XIV	2045	XIII
2028	XV	2046	XIV
2029	XVI	2047	XV
2030	XVII	2048	XVI
2031	XVIII	2049	XVII
2032	XIX	2050	XVIII

Another intriguing pattern is created by phases from one lunation to the next. The time of occurrence of a specific phase alternates from day to night. For example, if the exact time of a full moon falls during the day, in the next lunar cycle it will fall during the night. The same is true for each phase, alternating from day to night from one lunar month to another. Yet another recurring cycle shows up about every 15 lunations. Given a particular phase, the same phase will occur within about 1½ hours after 15 lunations have passed, a period of a year and a few weeks or approximately every 433 days.

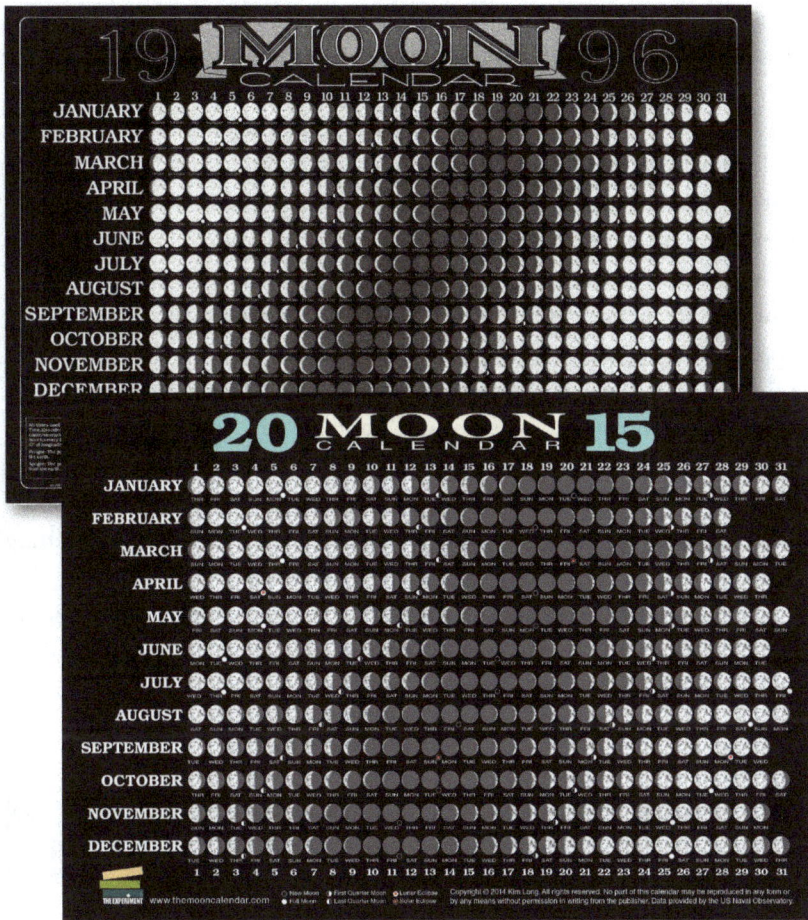

Two editions of the *Moon Calendar,* 19 years apart, illustrate the recurring pattern known as the Metonic Cycle. Because of variations in the number of calendar days between lunar phases—first quarter moon to full moon, or new moon to first quarter moon, for example—the 19-year pattern does not always repeat itself exactly.

MOON RISE AND MOON SET

The Moon rises and sets at different times every day because the civil calendar is based on a solar timetable, not a lunar one. On average, moon rise and moons et are about one hour later each succeeding day, but the time changes considerably from one location to another. Both latitude and longitude have an effect on this change. In some northern latitudes (northern Canada, for instance), there is a more dramatic change from day to day in the times for moon rise and moon set.

During some periods of the lunar cycle, there are days when the Moon does not rise or the opposite, does not set, as this event happens after midnight, and thus falls on the following day. During each lunar month, there is one day with no moon rise and one day with no moon set. This happens because the Moon "lags behind" the 24-hour day. The Moon actually has a 25-hour day (approximately). Therefore, for example, if the Moon sets at 11:50 P.M. on a Tuesday night, 25 hours later would run past Wednesday and the next setting time would be about 12:40 A.M. on Thursday morning.

The full moon closest to the fall equinox (in September) the Harvest Moon, rises at close to the same time from evening to evening, at least for a few days. This phenomenon adds to the amount of moonlight available, traditionally a boon to farmers working on fall harvest. (see page 124).

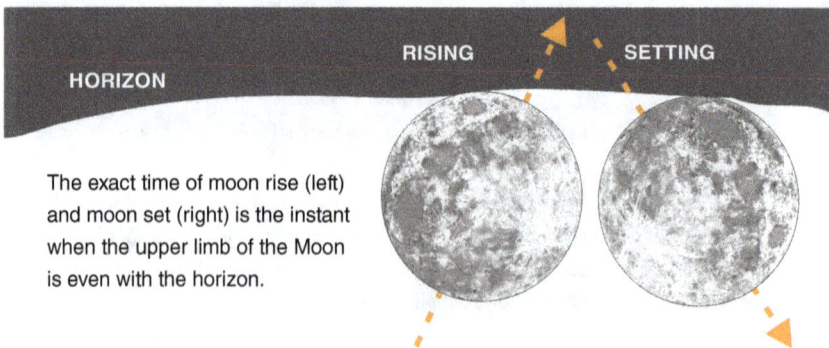

RISING   SETTING

HORIZON

The exact time of moon rise (left) and moon set (right) is the instant when the upper limb of the Moon is even with the horizon.

The time of moon rise is affected by latitude and longitude. This map shows the variation in times of moon rise for different locations in the eastern United States for October 24, 1999.

Buffalo, New York
5:40 PM

Albany, New York
5:22 PM

Detroit, Michigan
6:00 PM

New York, New York
5:25 PM

Philadelphia, Pennsylvania
5:30 PM

Pittsburgh, Pennsylvania
5:47 PM

Chicago, Illinois
5:17 PM

Indianapolis, Indiana
5:13 PM

Washington, D.C.
5:39 PM

Louisville, Kentucky
5:16 PM

St. Louis, Missouri
5:31 PM

Charlotte, North Carolina
5:55 PM

Memphis, Tennessee
5:34 PM

Atlanta, Georgia
6:14 PM

Birmingham, Alabama
5:22 PM

Jacksonville, Florida
6:05 PM

DAY-TO-DAY DIFFERENCES IN FULL MOON RISE

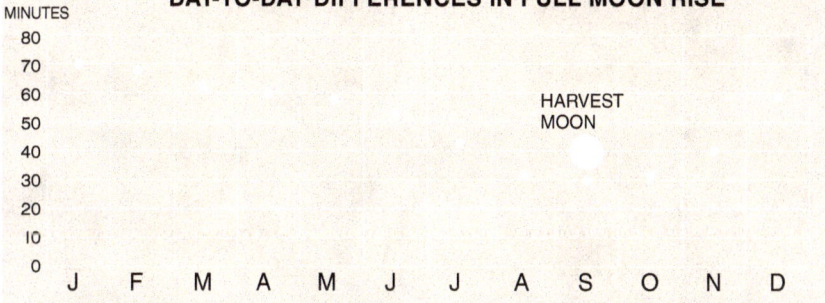

MINUTES

80
70
60
50
40
30
20
10
0

HARVEST
MOON

J F M A M J J A S O N D

At the time of the Harvest Moon, the full moon rises with the least difference in local time from night to night. This chart depicts a typical year in a mid-northern latitude in North America, with as little as 30 minutes difference from one day to the next. At the other extreme, in January, the lag is as much as 70 minutes.

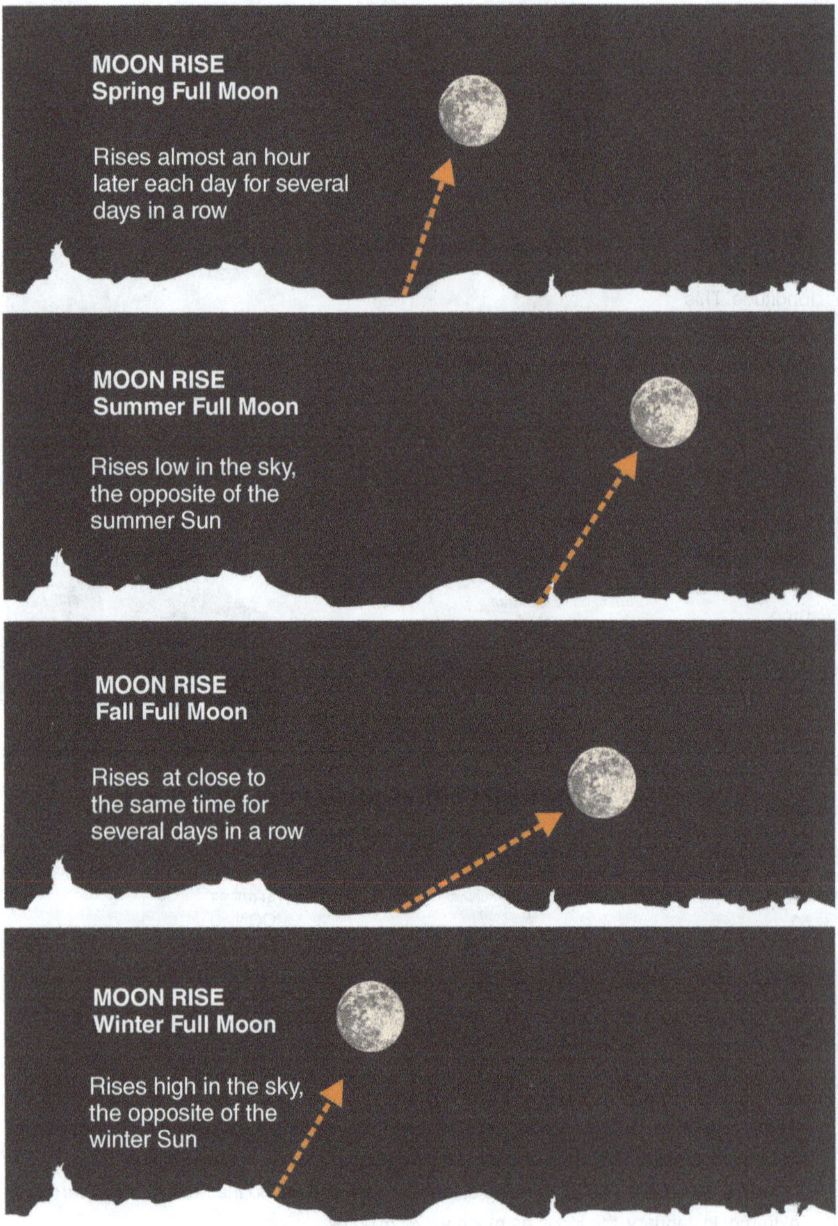

MOON RISE
Spring Full Moon

Rises almost an hour
later each day for several
days in a row

MOON RISE
Summer Full Moon

Rises low in the sky,
the opposite of the
summer Sun

MOON RISE
Fall Full Moon

Rises at close to
the same time for
several days in a row

MOON RISE
Winter Full Moon

Rises high in the sky,
the opposite of the
winter Sun

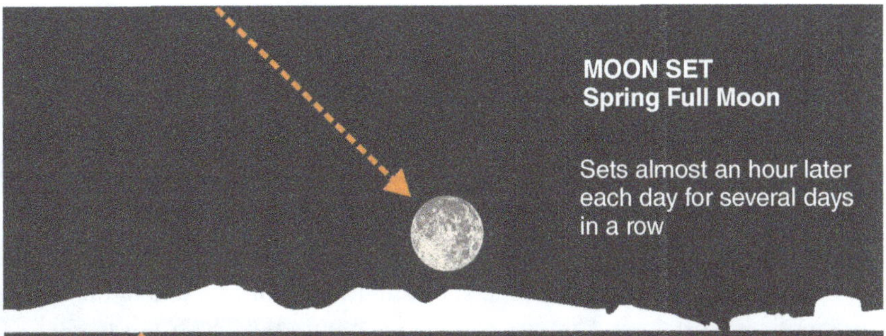

MOON SET
Spring Full Moon

Sets almost an hour later
each day for several days
in a row

MOON SET
Summer Full Moon

Sets low in the sky, the
opposite of the summer
Sun

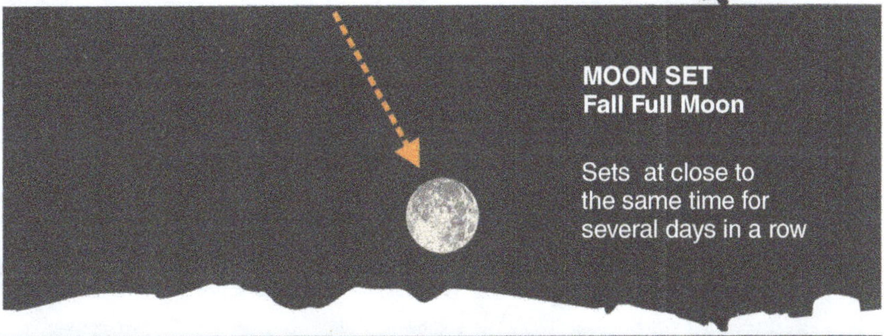

MOON SET
Fall Full Moon

Sets at close to
the same time for
several days in a row

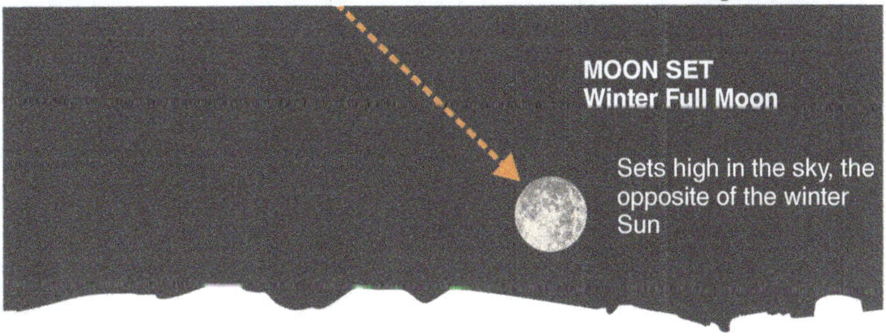

MOON SET
Winter Full Moon

Sets high in the sky, the
opposite of the winter
Sun

As the moon is rising and setting at different times for observers at different geographic locations, it is also rising and setting at different points on the horizon. From north to south, the Sun's path on the ecliptic at the horizon varies by a little less than 50 degrees. Because the Moon's path is roughly the same, this pushes the points of moon rise and set on the horizon to about the same variations — the Moon's path is tilted from the ecliptic by about 5 degrees, adding this additional distance when at its extreme.

During a period called the lunar standstill (a cycle of 18.6 years), the Moon rises and sets at the most extreme points on the horizon. This last happened over the period 2005–2007 and will happen again from 2023 to 2026.

		NEW MOON	FIRST QUARTER	FULL MOON	LAST QUARTER
WINTER	RISE	southeast	east	northheast	east
	SET	southwest	west	northwest	west
SPRING	RISE	east	northheast	east	southeast
	SET	west	northwest	west	southwest
SUMMER	RISE	northheast	east	southeast	east
	SET	northwest	west	southwest	west
FALL	RISE	east	southeast	east	northheast
	SET	west	southwest	west	northwest

MOON DIRECTIONS

As the Moon rises in the East, it follows a path across the sky that is close to the ecliptic—the path made by the Sun as it crosses the sky. When the Moon's phase is new, the Moon's path is closer to that of the Sun that at other times in its cycle, so in this period the position of the Sun in the sky approximates where the Moon is, or will be on the day of observation.

When the Moon is full it is opposite to the Sun and at the other end of the tilted ecliptic. Therefore, if the Sun appears high in the sky—characteristic of Summer months—the full moon will appear low; Winter is the opposite, when the Sun, appearing low in the sky, will be followed by a full moon that is high.

Height is represented in astronomical terms as declination, a measurement determined in degrees. The starting point is an imaginary circle projected out from the equator, at 0°, with 90° representing the highest point, the celestial pole, a projection of the axis of the Earth. Altitude is a related measurement, but uses the horizon as the starting point.

FULL MOON AT ITS HIGHEST POINT

Figures for a specific year in a middle latitude (39°) — the height in degrees for any full moon varies by location and date

66° JAN 23 · 60° FEB 22 · 50° MAR 23 · 44° APR 21 · 35° MAY 21 · 31° JUN 20 · 32° JUL 19 · 38° AUG 18 · 44° SEP 16 · 53° OCT 15 · 64° NOV 14 · 68° DEC 13

The Moon rises to its zenith, the highest point above an observer, at different times during each cycle. In traditional almanacs and other resources used for navigation, the Moon's zenith was referred to as its transit (other celestial bodies have their own transits), a point halfway between due east and due west. This was sometimes called its "southing," as the event occurs when the Moon is to the south. The transit point for the Moon is when it is at its highest point of a virtual line running through an observer from north to south.

As the Moon or any celestial object climbs in the sky, its height above the horizon is marked in degrees, from 0° to 90°. The highest point to which the full moon rises throughout North America is less than 90 degrees—only close to the equator does the Sun and Moon get so high in the sky that it reaches such an extreme altitude. In southern cities in North America—Miami, for example—the extreme height is about 89°; in Seattle it's 66°.

Moon rise and moon set at the horizon are noted with the term azimuth. Azimuth is measured in degrees at the horizon: the staring point is north (0°); east is 90°; south is 180°; west is 270°. In practical applications, this is roughly the same as compass readings. This is not the same thing as right ascension, however, which measures the angle to the west (right) of the vernal equinox along the plane of the celestial equator.

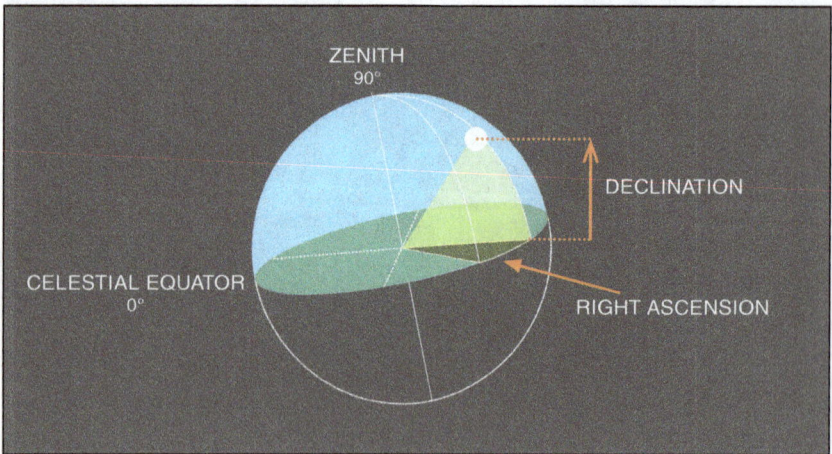

ZENITH
90°

DECLINATION

CELESTIAL EQUATOR
0°

RIGHT ASCENSION

MOON SIZE AND MOONLIGHT

The apparent size of the Moon to observers on Earth is about half of a degree. This is also the apparent size of the Sun. If you hold your hand up to the Moon with arm outstretched, this corresponds to about half the width of one finger—this ratio is fairly consistent no matter the size of the observer. Sometimes, however, the Moon appears to be much larger. When rising and setting, for example, the Moon at its full phase often appears to be larger than when it is directly overhead.

There are factors that can change the diameter and brightness of the Moon, but this apparent change in size is a phenomenon associated with rising and setting and is actually an optical illusion—the proximity of the horizon adds a remarkable degree of visual influence. The eye is tricked into measuring the Moon against nearby objects and the visual elements found on most horizons—buildings, trees, hills—create the impression of increased size.

This optical trick can be demonstrated by using cardboard masks to cut off the view of nearby objects. Make a mask with a viewing hole no larger than ¼ inch in diameter. Tape the mask to a yardstick so it can be kept at a fixed distance from your eye. As a full moon rises, view it through the mask when it just

Use your fingers and hand to approximate measurements at a distance. The little finger at the and of an outstetched arm equals about 1 degree (about twice the angular diameter of the Moon or Sun). The first and second fingers spread apart represent about 5 degrees and a fist, the width of the hand, is about 10 degrees.

1°   5°   10°

clears the horizon. Note the size of the Moon's disk compared to the size of the hole, then wait for the Moon to reach its zenith and view it again.

Although most of the apparent size change in the Moon is caused by visual trickery, there are physical effects which cause a measurable size difference, although not a large one. One of these effects is the difference in distance between an observer and the Moon from a horizontal perspective to a vertical one. This difference — the change from viewing the Moon on the horizon and at its zenith — is half the diameter of the Earth. However, this accounts for only a maximum difference of two percent in the diameter of the Moon, hardly enough to notice with the naked eye. And in any case, this size difference decreases the size of the Moon when it is at the horizon, not the opposite, because that is when it is farthest away from the observer.

A more significant difference is caused by the variation in the Moon's orbit around the Earth. This orbit is not round, but elliptical (oval) in shape, making some points in the orbit closer to the Earth than others. At the closest point (perigee), the Moon will appear measurably larger than when at the farthest point (apogee). The difference in diameter from one extreme to the other is about 10 percent.

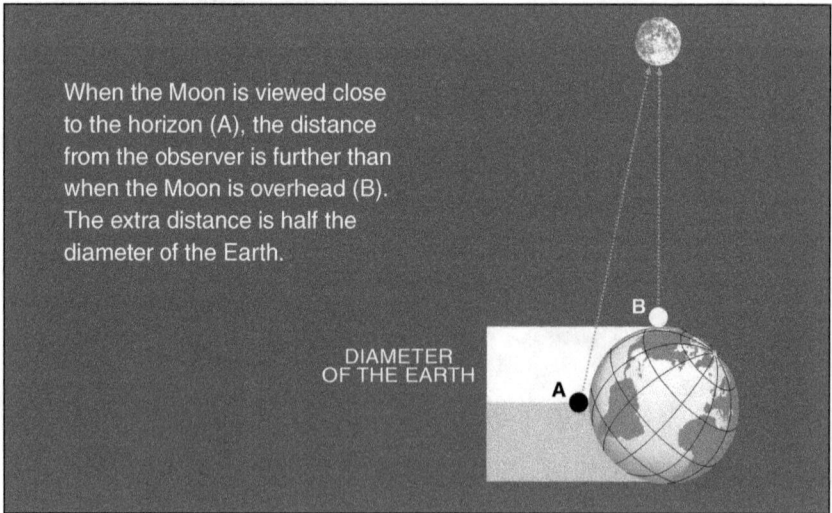

When the Moon is viewed close to the horizon (A), the distance from the observer is further than when the Moon is overhead (B). The extra distance is half the diameter of the Earth.

DIAMETER
OF THE EARTH

B

A

RELATIVE INTENSITY OF MOONLIGHT

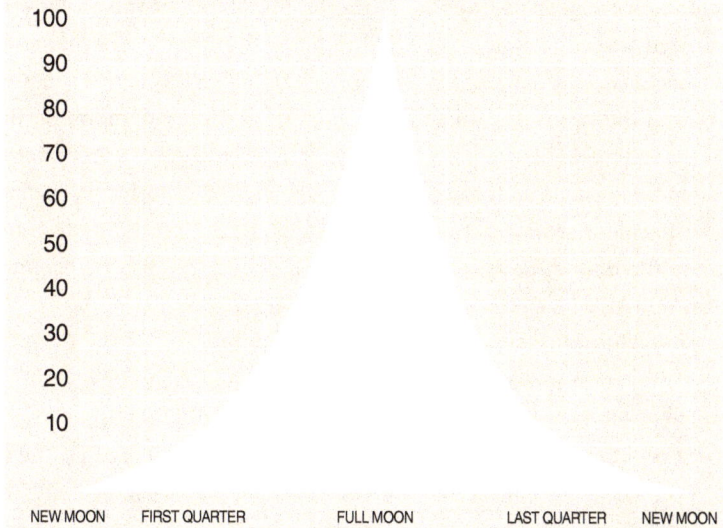

100
90
80
70
60
50
40
30
20
10

| NEW MOON | FIRST QUARTER | FULL MOON | LAST QUARTER | NEW MOON |

There is slightly more light from the waning crescent moon than the waxing crescent moon, but between the full moon and quarter phases, the waxing period is brighter than the waning period. Chart derived from Russell, 1916.

PHASE BRIGHTNESS

PHASE	AGE IN DAYS	PHASE ANGLE	RELATIVE BRIGHTNESS
New Moon	0.0	180°	0.00
	3.0	−143°	0.01
First Quarter	7.4	−90°	0.08
	10.0	−58°	0.22
Full Moon	14.8	0°	1.00
	18.0	39°	0.37
Last Quarter	22.1	90°	0.08
	25.0	125°	0.02

The Moon makes one orbit every 29.5 days, so there is one perigee and one apogee during every orbital period. However, the phases of the Moon are not synchronized with this cycle, and a full moon—the most visible phase—will typically occur on or near a perigee or apogee in only a few months in any calendar year. If the two cycles coincided in August, with the full moon occurring on about the same day as the perigee—give or take a few days—they might also coincide in September, with the date of perigee gradually "drifting" backwards through the calendar month faster than the date of full moon. By October or November, the date of perigee would be more than a few days before the full moon, and the distance effect would thus be diminished.

The same kind of distance factors that change the visible size of the Moon can affect the amount of sunlight it reflects. Light is affected by the law of inverse squares—it decreases as the square of the distance. This happens because light spreads out as it gains distance; the farther from the light source, the more area that the same amount of light covers. For example, if you double the distance between you and a light bulb from 10 to 20 feet, the illumination would drop by 75 percent.

The Moon does not alter its distance from the Sun by much, but there is enough change to affect the intensity of light it receives. Also, the changing distance between the Moon and Earth affects the amount of sunlight reflected from the Moon to Earth. However, these changing values are rarely noticeable because of the interfering effects of the Earth's atmosphere, which diffuses the moonlight.

The first quarter moon and the last quarter moon are only about 8 or 9 percent as bright as the full moon.

As light travels farther from its source, it spreads out, covering more area but losing intensity. This phenomenon is ruled by the inverse square law, where the intensity of the light is inversely proportional to the square of the distance from the light source.

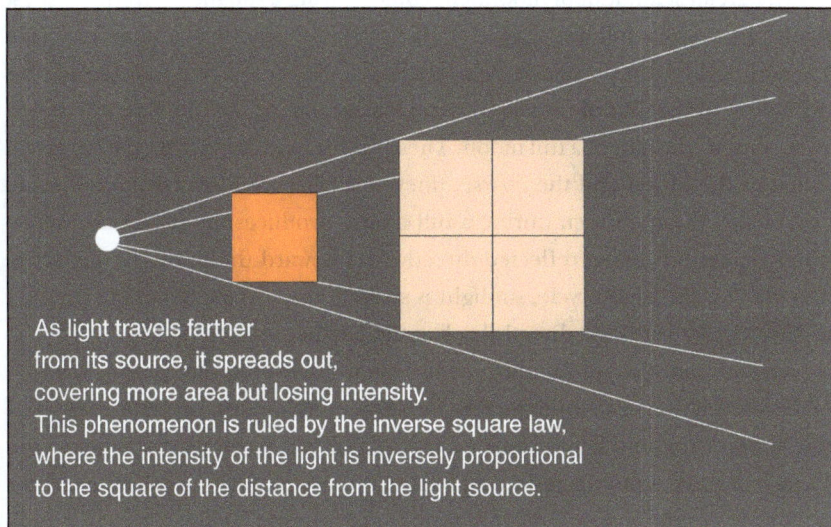

MOON BRIGHTNESS
RULES OF THUMB

- Full moons vary in brightness by about 10 percent (not including factors such as light pollution and atmospheric conditions).

- The Moon is not larger as it rises and sets, but it looks bigger because of the visual effect created by the closeness of the horizon.

- The Moon appears larger when it rises in the fall and sets in the spring because it rises and sets at a shallower angle at these times of the year, placing it closer to the horizon.

- The Moon appears larger when it is closest to the Earth (perigee). This happens once every lunar month, and when the perigee occurs close to the time of full moon, it has the greatest visual effect.

The greatest amount of moonlight comes during a full moon. At the highest point in the sky, a full moon yields 0.01 to 0.03 foot-candles of illumination, a fraction of the light produced by a candle at a distance of one foot (hence the derivation of the unit of measurement. Quarter moons provide only about 8 to 9 percent of the light of a full moon. These differences come from the spherical shape of the Moon and the coarse, uneven surface. Sunlight striking the full face of the Moon, as seen during a full moon, produces the maximum reflection because it is being reflected directly back toward the Earth. At any other time during the phase cycle, sunlight is striking the surface obliquely, limiting its ability to be reflected directly back at the Earth.

Recently, another unusual factor has been discovered that plays a key role in the production of moonlight. Examination of moon dust brought back from the Apollo expeditions has revealed the role of tiny particles that cling to the surface of lunar sand. In reflecting sunlight, these particles act to amplify rays of light, a condition labeled "coherent backscattering." Under certain conditions, such as during a full moon, the reflection intensifies, producing more visible light than during other phases.

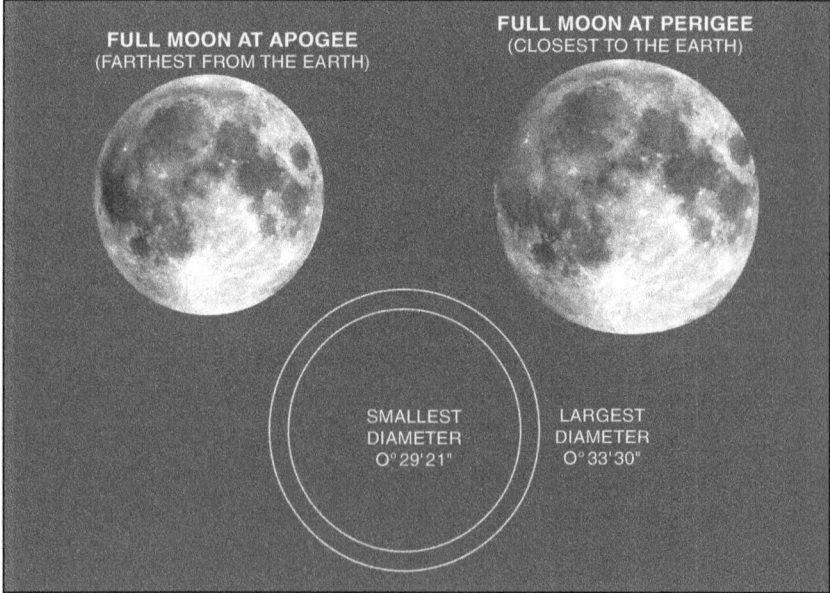

FULL MOON AT APOGEE
(FARTHEST FROM THE EARTH)

FULL MOON AT PERIGEE
(CLOSEST TO THE EARTH)

SMALLEST
DIAMETER
0° 29'21"

LARGEST
DIAMETER
0° 33'30"

LUNAR ECLIPSES

Lunar eclipses are produced when the Earth's shadow falls on the Moon. This would happen every full moon if the Moon orbited around the Earth in the same plane as the Earth orbits around the Sun. The path of the Sun, called the ecliptic, got its name because of this potential. The Moon's orbit, however, is tilted about 5 degrees above the Earth-Sun plane. This tilt itself, however, rotates, generating eclipses only when the tilt of this plane crosses the ecliptic at the right time, just when the Moon is new or full.

A lunar eclipse is visible over an entire hemisphere and is seen at the same time to everyone who is in sight of the full moon. Lunar eclipses can last for more than three hours because the Moon and the Earth are moving slowly in relation to each other, and the shadow cast by the Earth is so large. Because of their sizes and the relative distances between the Earth, Moon, and Sun, this shadow is much larger than that cast by the Moon on the Earth during a solar eclipse, which, correspondingly, is much shorter.

Although eclipses are always caused by the same general lineup of Sun, Moon, and Earth, each lunar eclipse may have its own unique visual characteristic. Colors and the deepness of the shadow on the surface are affected by the type of eclipse, local weather conditions, atmospheric conditions, and the geographic location of the observer. Volcanic activity may also affect the visibility of an eclipsing Moon, usually making it darker and less visible; sometimes the volcanic dust in the upper atmosphere may create unusual colors. During a total eclipse, the appearance can also be affected by solar conditions, particularly sunspot activity and the relative distance between the Moon and the Sun. A relationship is known to exist, for example, between the 11-year cycle of solar activity and the brightness of lunar eclipses, with the eclipsed moons

UMBRA   PENUMBRA

The shadow cast during an eclipse has two components, a darker central area (the umbra), and a lighter outer area (the penumbra).

37

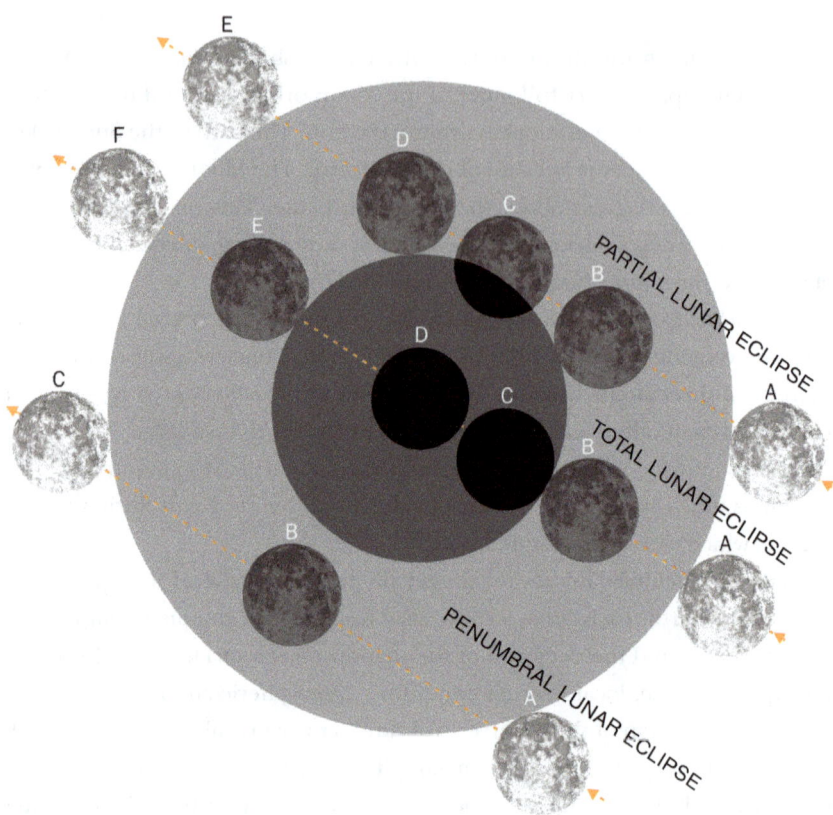

LUNAR ECLIPSE SEQUENCE

PARTIAL ECLIPSE

A Moon enter penumbra

B Moon enters umbra

C Middle of eclipse

D Moon leaves umbra

E Moon leaves penumbra

TOTAL ECLIPSE

A Moon enters penumbra

B Moon enters umbra

C Total eclipse begins

D Moon leaves umbra

E Moon leaves penumbra

PENUMBRAL ECLIPSE

A Moon enters penumbra

B Moon enters umbra

C Moon leaves penumbra

dimmer when solar activity is low. For total lunar eclipses, some astronomers use a standardized scale, known as a Danjon lunar eclipse scale, to accurately rate the Moon's appearance during these events.

There are three types of lunar eclipses.

PENUMBRAL ECLIPSE

This is a partial eclipse that occurs when the Moon only passes through the secondary shadow (penumbra) of the Earth. A penumbral eclipse is sometimes called an appulse eclipse. During a penumbral eclipse, the Moon's light is dimmed, but it does not go completely dark because the penumbral shadow is not deep enough to block out all of the Sun's illumination. Often, there is no visible line separating the shadow from the sunlight on the Moon's surface, and the eclipse is only noticeable as a slight darkening of the lunar surface. In a few rare penumbral eclipses, the Moon may only graze the edge of the penumbra, and observers may see little change.

DANJON LUNAR ECLIPSE BRIGHTNESS SCALE

André-Louis Danjon, a French astronomer, developed a method of rating the illumination of the Moon during lunar eclipses—the illumination comes from earthshine.

0 Very dark. At the middle of totality, darkness almost completely obscures the Moon.

1 Gray or brownish color. Moon is darkly shadowed with some features faintly visible.

2 Rusty brown or dark red color. Umbra is very dark in the center and lighter at the edges.

3 Reddish to brick red color. Edges of the umbra are lighter in color and may appear yellowish.

4 Orange or copper color. Edges of the umbra are very light and may appear bluish in color.

In a penumbral eclipse—and the penumbral phase of a total lunar eclipse, the shadow cast on the full moon dims its illumination, but not enough to obscure its surface features completely.

TOTAL ECLIPSE

The Moon passes completely through the main shadow (umbra) of the Earth. The dark umbral shadow cast by the Earth does not completely obscure the Moon but changes its color to a dull copper tone, an effect created by earth-shine (light reflected off the Earth onto the Moon, see page 46); this reflected light illuminates features of the Moon's surface. The color is created by the filtering effect of the Earth's atmosphere, which removes all but the red wavelengths of sunlight. The Moon can stay in the umbral shadow of the Earth for as long as 90 minutes, but the movement through the penumbral shadow can last for about 60 minutes.

In a total lunar eclipse, the umbral shadow is dark enough to almost completely obscure the surface features of the Moon, but earthshine, with a reddish tint, adds its own illumination.

PARTIAL ECLIPSE

The Moon partially enters the main shadow (umbra) of the Earth. These eclipses do not produce the reddish colors characterized by total lunar eclipses, because the secondary shadow is not deep enough to highlight the reflected light from the Earth.

ECLIPSE RULES OF THUMB

- Full moons are the only time lunar eclipses occur.

- New moons are the only time solar eclipses occur.

- A solar eclipse always occurs two weeks after or two weeks before a total lunar eclipse.

- Lunar eclipses can last for a maximum of 3 hours and 40 minutes, with the period of totality lasting for as long as 1 hour and 40 minutes.

- Solar eclipses can last for a maximum of 7 minutes and 40 seconds if they are total (at the equator), 12 minutes and 24 seconds at most if they are annular.

- Lunar eclipses can never happen more than three times a year. Solar eclipses happen at least twice a year but never more than five times a year.

- Lunar eclipses are visible over an entire hemisphere. Solar eclipses are visible in a narrow path that is a maximum of 167 miles wide (269 km).

- The greatest number of solar and lunar eclipses that can happen in a year is seven.

- At any specific geographic location on the globe, a total solar eclipse can occur only once every 360 years, on average.

- Solar eclipses and lunar eclipses go together in pairs. A solar eclipse is always followed or preceded by a lunar eclipse, within an interval of 14 days. Eclipses may also occur in threes, alternating lunar, solar, lunar.

- The characteristics of one eclipse are repeated every 18 years, 1 day, and 8 hours, with some minor variations. This long-term rhythm is called the Saros cycle. At any given time, there may be several dozen different versions of this cycle in effect.

SOLAR ECLIPSES

The Moon also causes eclipses of the Sun. When the Moon comes direct-ly between the Earth and Sun — this can only happen during the new moon phase — it blocks out the Sun's rays. Depending on how the Moon and the Sun line up, there can be either partial or total blockage of the Sun's disk. In most months, the Moon's path is too high or too low as it crosses the Sun's position during the new moon — the two bodies are close enough that observers can't see the Moon for at least a day, but no eclipse occurs.

A total solar eclipse occurs when the Moon completely blocks out the Sun. However, the elliptical orbit of the Moon can place it at varying distances from the Earth when such an event occurs. When the orbit is closer to the Earth during a solar eclipse, the Moon appears larger from the perspective of Earth and therefore blocks out the Sun for a longer period of time. When the orbit is farther from the Earth during an eclipse, the Moon appears smaller and may not even completely cover the Sun's disk, allowing a thin ring of light around the eclipse shadow. This event is referred to as an annular eclipse.

A solar eclipse is much shorter in duration than a lunar eclipse because the Moon's shadow is falling on a rapidly rotating Earth. The maximum time for a solar eclipse is 7 minutes and 40 seconds (if it is located on the equator) but most are much shorter. The moving shadow cast by the eclipse is called an eclipse track and is usually about 3,000 miles long and 100 miles wide, but the width can vary from almost nothing to a maximum of 167 miles wide. Although observers may see some of the effect of a solar eclipse from a large geographic area, only observers in the direct track of totality will witness its complete effect.

The Moon's shadow during a solar eclipse moves across the Earth's surface at speeds between 1,000 and 2,000 miles per hour. The slowest movement is at the equator and the fastest at the poles. The speed of the shadow is caused by the combined movements of the Moon and the Earth; the Moon is moving at about 2,000 miles per hour eastward and the Earth (at the equator) rotating at about 1,000 miles per hour.

In any kind of solar eclipse, viewing should only be done with appropriate equipment to avoid damage to the eyes.

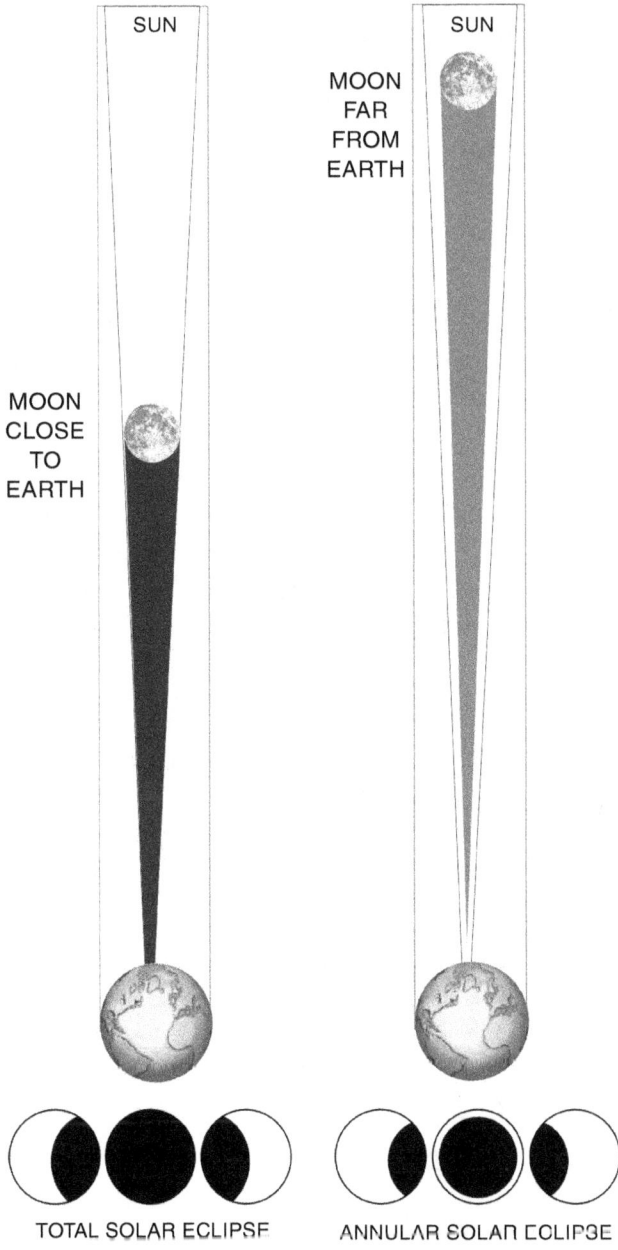

SUN

MOON
FAR
FROM
EARTH

SUN

MOON
CLOSE
TO
EARTH

TOTAL SOLAR ECLIPSE

ANNULAR SOLAR ECLIPSE

There are three kinds of eclipses of the Sun.

PARTIAL ECLIPSE

The Moon is in conjunction with the Sun but its path does not take it directly across the center of the Sun's disc. The visible result is a dark blob obscuring part of the Sun's face.

ANNULAR ECLIPSE

The Moon is in conjunction with the Sun, but is at a far point in its orbit around the Earth. The Moon's image is thus too small to completely cover the Sun's disc, and a ring of sunlight is visible around the edges.

TOTAL ECLIPSE

The Moon is in conjunction with the Sun and is at a close point in its orbit around the Earth. The Moon's image is large enough to completely block out the Sun's disc; the visibility of the total eclipse extends over a long track, but for only a relatively narrow width.

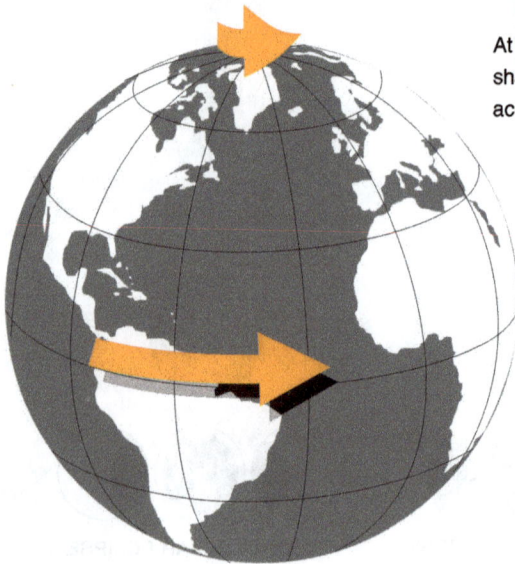

At the poles, an eclipse shadow moves at 2,000 mph across the Earth's surface.

At the equator, an eclipse shadow moves at 1,000 mph across the Earth's surface.

TOTAL SOLAR ECLIPSE OF 1999 AUGUST 11

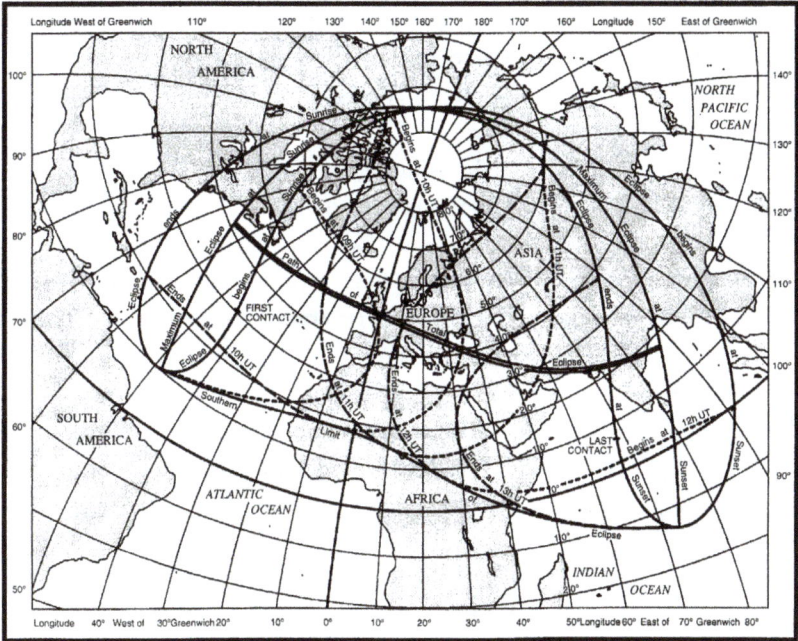

Several key resources provide accurate, reliable information about lunar and solar eclipese, often years in advance. Maps such as the one above give useful information about where each eclipse will be visible, and in the case of solar eclipses, the specific geographic tracks of the totality.

Annular eclipse viewed from south of Santa Fe, New Mexico, on 5-20-2012.

EARTHSHINE

The light of the Sun reflects not only off the face of the Moon, it also reflects off the Earth's surface. When the reflected sunlight from the Earth produces visible light on the Moon, it is referred to as earthshine. During periods when the Moon is lit fully by sunlight—during the full moon—earthshine is not produced because the Sun is in opposition, the wrong position to produce this effect, which would not be visible in any case because of the brightness of the direct illumination. During most of the lunar month and during eclipses, however, earthshine may be seen if atmospheric conditions permit. It is visible as a dark, grayish, dull copper, or reddish hue. During total lunar eclipses, this phenomenon is most visible.

Just after the new moon, the emerging first crescent moon sometimes produces an effect referred to as the "old moon in the new moon's arms." This phenomenon is visible just after sunset a few days after the new moon when the young crescent moon is beginning to set in the west. A thin crescent-shaped illumination on the lower right side of the Moon's face will be seen; the rest of the surface may also stand out from the surrounding sky as earthshine, with a copper or reddish hue.

For early risers, the same effect is sometimes visible as the old crescent moon just before sunrise. The best time of the year to experience this phenomenon is in late winter and early spring.

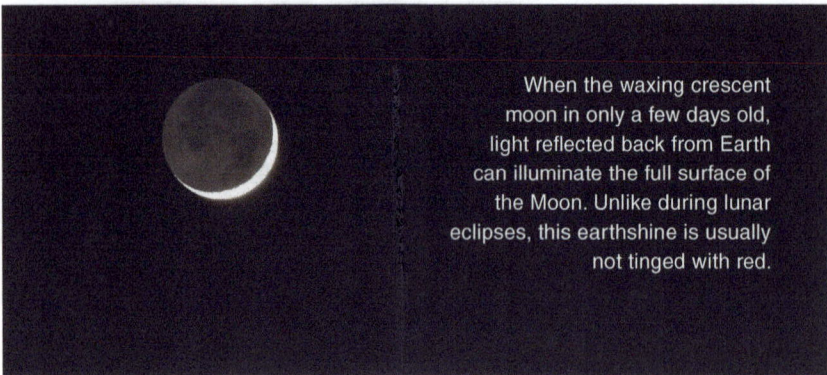

When the waxing crescent moon in only a few days old, light reflected back from Earth can illuminate the full surface of the Moon. Unlike during lunar eclipses, this earthshine is usually not tinged with red.

MOONLIGHT EFFECTS

A strong source of light such as the Sun or Moon can produce interesting optical effects when combined with the right atmospheric conditions. A rainbow—produced when sunlight is refracted through water droplets in the air—is perhaps the most familiar of these effects. When the light from the Moon is refracted through water droplets, a similar effect can be produced. This prismatic display is known as a moonbow or lunar rainbow, but is much less intense than a solar rainbow and observers may see only the palest of colors. Moonbows are most likely to be seen when the Moon is full or within a few days of full.

When moisture is present high in the atmosphere in the form of cirrus clouds, it is frozen into ice crystals. If the Moon is in the right position, moonlight can form a halo or ring around itself. Ice crystals have six sides and normally refract light at an angle of 22 degrees, typically creating a lunar halo that is 22 degrees in diameter. Less typically, the halo may be produced by different angles in the crystals and appear 46 degrees in diameter. The same action that produces this ring also breaks the light into colors, just like a prism. The inner edge of the halo is blue and the outer is red when it is 22 degrees; the colors are reversed when it is 46 degrees.

Moon dogs, also known as mock moons, are also produced by the interaction of moonlight and moisture in the atmosphere. With the right combination of humidity and angle, an observer may see a paler halo-like image of the Moon off to the side of the Moon itself. The official name for this effect is a paraselene if the extra image is seen 22 or 46 degrees away. If the image is at 90, 120, or 140 degrees, it is called a parantiselene. If it is at 180 degrees, it's an antiselene.

Moon pillars are another form of halo related to the Moon, but are rarely visible. Moon pillars can be seen when the Moon is near to the horizon, either just before or just after setting or rising. These are pale shafts of light above or below the Moon and are caused by moonlight reflecting off of ice particles or snowflakes in the atmosphere.

A corona is another type of light show created by the Moon. Similar to a lunar halo, a corona is produced by high, thin clouds, but it is not as large in diameter as a halo and often appears as a diffuse glow. A typical lunar corona is only one or two degrees in diameter and closely fringes the Moon itself, sometimes passing in front of the Moon when it is behind a layer of clouds. A corona also may appear in several colors, like a solar rainbow, although the colors are not as intense. Typically, a slight reddish or bluish tint is visible. On rare occasions, two or more coronas in a concentric pattern may be produced at the same time. A lunar corona can also appear along with a lunar halo.

Moon coronas can be produced by a layer of visible clouds, sometimes producing one or more colors.

A lunar halo, also known as a "ring around the Moon," is caused by light refracted through ice crystals in cirrus clouds. The six sides of these ice crystals refract the light at a 22 degree angle, almost always producing a halo that is 22 degrees in diameter. This is a graphic depitction of a lunar halo, not a photograph.

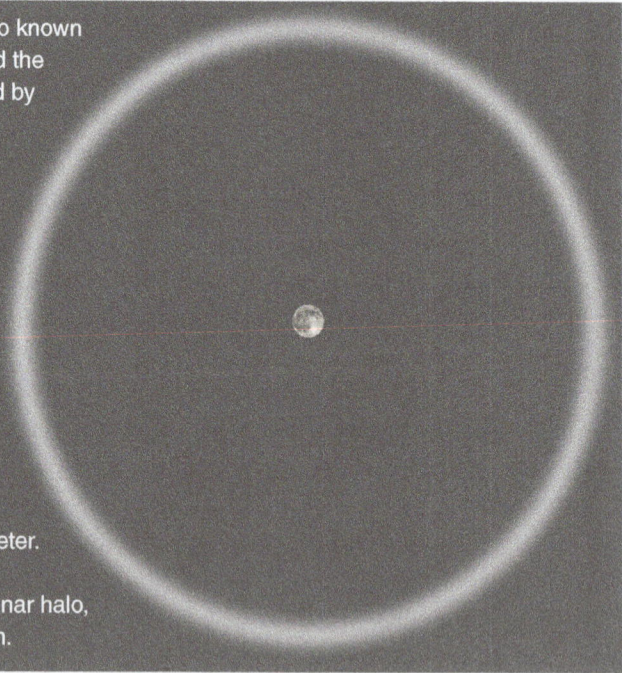

MOON BOUNCE

Although far from Earth, the large area represented by the surface of the Moon makes a tempting target for radio signals. Amateur radio operators (aka "hams") around the world occasionally use the Moon for "moon bouncing," or Earth-Moon-Earth communications (EME). Most hams do not have equipment strong enough for this feat, but an estimated 1,000 are able to participate.

Military forces were the first to use the Moon for this purpose, beginning in the 1950s. The first successful transmission was on July 24, 1954, with a voice signal sent from a facility at the Naval Research Laboratory in Maryland; it returned about two and a half seconds later. The project was designated "Operation Moon Bounce." Earlier unsuccessful attempts are thought to date to at least 1928; a radar echo was first detected in 1946 during an experiment called Operation Diana.

The first successful moon bounce attempt by amateurs was in 1964, when radio operators communicated between Stanford, California, and Australia. Although the Moon's surface is relatively easy to aim at, ham hobbyists note that the uneven surface, atmospheric conditions on Earth, and the distance between the two bodies makes it a tricky operation. Because of the distance the radio signals must travel, it can take several seconds for messages to be transmitted and received.

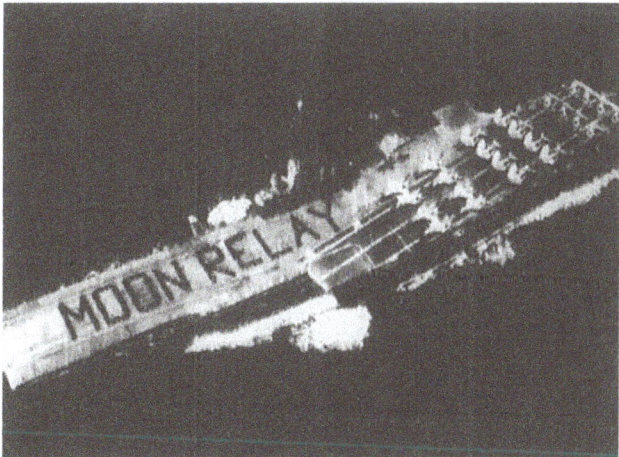

The first image transmitted to and from the Moon was this faxed photo, of the USS Hancock, on January 28, 1960. It was sent from Honolulu to Washington, DC.

Photo courtesy of the Naval Research Laboratory.

THE MOON IS FLAT?

Early observers of the Moon, including Galileo, noticed that the full moon appeared to be flat. If a round object such as a balloon or ball is illuminated like the full moon, a distinct three-dimensional effect occurs. In fact, almost any sphere exhibits this common effect from illumination—a gradual darkening around the edges with the brightest area in the center. The Moon does not produce this phenomenon.

That the full moon has almost equal illumination everywhere across its surface is the result of its unusual surface texture. Light from the Sun striking the Moon is almost completely absorbed by the surface. Only about 7 percent of the sunlight is reflected. The percent of light reflected from an object is referred to as albedo, designated as a measurement from 0 to 1, with 1 being 100 percent reflection. The Moon's albedo is 0.07. In comparison, the albedo of Earth is about 0.30 (30 percent); Venus is 0.56; and for Neptune it's 0.73.

The texture and color of different substances affects the albedo. For instance, on Earth the albedo of concrete ranges from 0.17 to 0.27; the albedo of snow is 0.45 to 0.90; the albedo of deserts is 0.25 to 0.30; and the albedo of soil is 0.05 to 0.15. Various features on the Moon also have different albedos. This results from the primary materials present at each location. The range is from 0.05 (Sinus Medii) to 0.176 (Aristarchus). Even during eclipses, the lunar features with the highest albedos can be seen because of earthshine.

The almost complete absorption of light on the Moon is caused by the rough, uneven texture of the surface and a layer of material made of particles less than a half inch in diameter. The particles of this lunar soil are themselves coated with a fine layer of rock powder that effectively scatters the incoming sunlight. Around the edges of the illuminated disk of the full moon, this redirection of light almost eliminates the shadows that would ordinarily be seen on the sides of a typical spherical object, effects that produce the visual image of roundness.

MOON EFFECTS

Because of its prominence in the sky, cultures throughout the world—ancient and modern—have attributed special powers to the Moon. Religion and astrology aside, the gravitational power and light source represented by the Moon do have measurable effects on Earth.

Tides are the most obvious of these, with the ebbing and flowing being a key factor associated with general feeding activity in water-dwelling animals. Filter feeders such as oysters, for one, increase their feeding activity during high tide, taking advantage of the additional food material swept in by tidal currents.

Moonlight also affects activities on Earth. Zooplankton, for one, have been found to closely follow lunar phases, especially in Winter months, with large populations of these important tiny animals migrating up and down columns of water in response to the varying degrees of illumination coming from the Moon. Scientists believe this migration pattern is to reduce predation; evidence of this cycle has been noted throughout the world's oceans.

Predators such as lions are known to be more active predators during the week just after a full moon. This period, with an increasing period of darkness just after the Sun sets, provides a lack of visibility of prey and the lions are thought to be hungrier because just before this period, nights lit by the full moon reduced their hunting effectiveness, making them hungrier. Around the globe, in fact, many species of prey animals reduce their activity during the nights immediately around full moons, and are more active on other nights during the lunar cycle, a direct effect of the nighttime illumination.

Human activity has been popularly linked to the full moon cycle as well, though numerous studies—including emergency room visits, births, accidents, sleep patterns, the stock market, and other measurable concepts—have failed to establish a link. Weather has also long been linked to lunar cycles, folklore that persists today, though extensive studies of data do not support this connection, though recent computer models hint at the potential of minor correlations with the Jet Stream. Traditional gardening activities—planting, weeding, harvesting, etc.—are also commonly tied to lunar phases, but no scientific studies have been able to show a correlation.

FIRST SIGHTING

Every lunar month begins with the phase of the new moon. This phase, however, provides little interest for observers unless there is an eclipse because there is nothing to see. By definition, the Moon is too close to the Sun to see. Within one or two days, however, the first crescent moon appears in the western sky just after the Sun sets, providing a striking visual event.

For some sky watchers and amateur astronomers, much effort and time is spent attempting to sight the first crescent moon at the earliest possible time after the new moon. This task is not as easy as it seems, as visibility of the slender, illuminated crescent is affected by a number of variable factors, including local humidity (which produces haze in the atmosphere), weather conditions, the time of year, the geographical location of the viewer, and the time of day when the new moon occurs.

The current record for the earliest crescent moon spotted with the naked eye is 15 hours and 32 minutes. Using binoculars, a first crescent moon has been seen at 13 hours, 32 minutes. With a telescope, the earliest sighting is 11 hourss and 40 minutes after the new moon.

Ordinarily, the first crescent moon will be easily visible within three calendar days of the new moon, and in most months, the first crescent can be sighted by the second calendar day after the occurrence of this phase. But in order to spot the emerging crescent sooner, extra care and planning is required. Astronomers calculate that the in order for the first crescent to be theoretically visible, the Moon has to be at least 7.0 to 7.5 degrees away from the Sun, although some sources believe the minimum to be about 10 degrees and others suggesting it is 13 degrees. It is not simply a matter of hours after the new moon, but the elongation—angle of separation from the Sun—because depending where it is in its orbit—apogee or perigee—the Moon is moving at different speeds, and the number of hours it takes to get to 7 or 10 degrees may vary.

The major problem associated with seeing the crescent moon is its lack of illumination. Because so little of the surface of the Moon is lit and the illuminated crescent is backlit by a sky that is not yet dark, it provides less of a visual target for the human eye. This illumination is produced by the nearness of the Moon to the Sun; it is just leaving the point in time of the new moon, when it is as close to the Sun as it gets every month. This means that during this period—from a few days before the new moon to a few days after the new moon—the Moon's position in the sky is close to the Sun and flooded with its light.

RECORD SIGHTINGS

The earliest that first crescent moons have been sighted:

Naked eye	15 hours, 32 minutes
Binoculars	13 hours, 32 minutes
Telescope	11 hours, 40 minutes

The illumination of each crescent moon varies from month to month because of the varying angle between the Moon and the Sun; this is called the elongation or arc of light. When this angle is zero degrees, the Moon passes in front of the Sun, producing an eclipse, but most months do not produce eclipses because the angle is greater than zero. In non-eclipse months, elongation varies from one to five degrees at the time of new moon. The potential for sighting a first crescent moon is highest when the angle is greatest.

The illuminated area of the Moon is referred to as the bright limb, the unlit portion is called the dark limb, and the line separating the two is called the terminator. As the bright limb appears in the young crescent moon, it is a very thin slice along the edge pointing to the Sun—the extreme ends are called the cusps. Because of the irregular surface of the Moon, for several days as the crescent moon waxes, the cusps do not extend completely to the Moon's north and south poles.

LEFT First crescent moon at 27 hours. RIGHT First crescent moon at 36 hours.

Another variable that affects the amount of illumination is the apparent size of the Moon, which varies according to how far away from Earth it lies. This distance changes over time because of the elliptical shape of the Moon's orbit; when the Moon is closest to the Earth—at perigee—it is largest, providing a bigger target for observers.

The same conditions that produce a first crescent moon are duplicated for an old last crescent moon. During this phase, the Moon is gradually being overtaken by the Sun and creeps closer and closer to it from day to day. Instead of being an evening phenomenon, it is visible in the morning just before the Sun rises. And instead of being up and to the left of the Sun, as seen from Earth, it is up and to the right (for observers in the Northern Hemisphere).

Based on astronomical calculations and simulated perfect conditions, the earliest that a crescent moon could be seen from the earth is when there is more than 7 degrees of elongation. Elongation is the angle between the Sun and the Moon and is related to the age of the Moon and the tilted plane of the Moon's orbit relative to the Sun. At an extreme, the Moon may be as much as 5½ degrees from the Sun at the instant of the new moon; in order for a first crescent to be visible, it must be at least a few degrees more. Even at the instant of new moon, without the Earth's atmosphere spreading and deflecting the

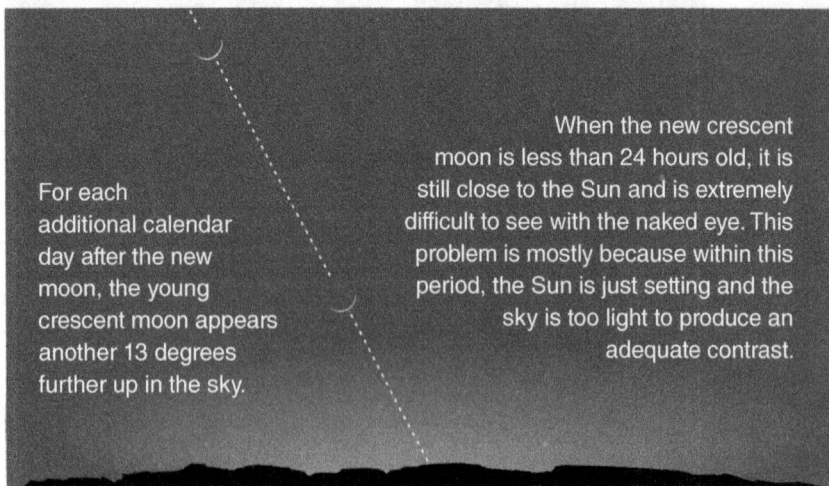

For each additional calendar day after the new moon, the young crescent moon appears another 13 degrees further up in the sky.

When the new crescent moon is less than 24 hours old, it is still close to the Sun and is extremely difficult to see with the naked eye. This problem is mostly because within this period, the Sun is just setting and the sky is too light to produce an adequate contrast.

FIRST CRESCENT MOON SIGHTING GUIDELINES

Moon's altitude	At least 10 degrees above the horizon at sunset.
Moon's elongation	At least 12 degrees away from the Sun.
Sunset	At least 10 minutes before the Moon sets.
Season	Late winter through early spring, when the ecliptic is at its steepest angle to the horizon.
Lunar month	New moon close to perigee.
Sky conditions	Seasons when the humidity is low to reduce haze—winter months are best. Regions such as the southwest are usually better for sightings because of a climate with low humidity.
Location	Site lines with the greatest unobstructed views of the western horizon. Higher elevations can improve site lines, looking down to the west.
Latitude	Lower latitudes provide better viewing options than higher latitudes.
Eyesight	The more acute your eyesight, the better the chances for sighting.

sunlight, it is theoretically possible to see the shape of the Moon because of the light reflected onto its surface from the Earth.

In fact, this was accomplished in 1966 when a photograph was taken from a special camera mounted in a rocket launched from the White Sands Missile Range in New Mexico. The photograph, taken when the Moon was only 2 degrees from the Sun, showed a small, irregular illumination on the side of the Moon facing the Sun.

For Earth-bound viewers, however, the quest to sight a first crescent moon existed before space flight and even long before telescopes. Many ancient calendars, in fact, because they were based on the movements of the Moon, human observation was required to keep track of the beginning of lunar months.

In the modern world, a few calendars are still based on lunar rhythms. Fundamentalist Muslims, for example, follow the traditional Islamic calendar, and must sight the first crescent moon in order to mark the beginning of calendar months. This sighting must be done without the aid of magnification and two reliable witnesses to each sighting are required in order for the new month to begin. Even with the aid of sophisticated computer programs that can closely predict when this event might happen, a physical sighting is still needed.

Depending on the country and the religious order, sightings of the first crescent moon are regulated in a variety of ways. The use of the unaided human eye is one constant, but use of computer programs, GPS equipment, and other modern technology may be permitted. In some countries—the United States is one—a centralized Islamic center is authorized to announce the sighting, and increasingly, Muslim countries have shifted to a formalized system of dates, announced in advance, to reduce the confusion that can come from competing sightings that differ by a day or more.

At its earliest stages, the first crescent moon is a striking sight. If the Moon were smoothly spherical, the lit area would extend close to one-half the distance around the circumference, or 180 degrees. But because the surface is rough and irregular, the arc that is illuminated will be somewhat shorter, typically about 130 degrees but can be as short as 30 degrees when the crescent is extremely new. And this narrow sliver of surface is not an even, symmetrical "slice," but dotted and broken along its inner edge from the effects of shadows of craters, valleys, and mountains.

Close to the ends of the arc, the cusps are also not perfect, but ragged and broken, further irregularity caused by the lunar landscape. Here, high elevations, such as mountain tops, catch the sunlight and form bright dots and small splotches amid darker areas.

When very young, the new crescent moon is silhouetted against a sky that is not yet dark. This diminishes another notable lunar effect, earthshine. Sunlight reflected off the Earth's surface and back to the Moon ordinarily strikes the complete visible surface of the Moon, even that part not illuminated directly by the Sun. The effect produces a distinct, although dim, outline of the lunar sphere. Near the horizon and viewed against a bright sky, as when the first crescent moon occurs, earthshine is less likely to be visible and may not appear at all if the crescent moon is less than 24 hours old.

OBSERVING THE FULL MOON

The full moon occurs at a precise moment in time. In astronomical terms, it is full at the moment when it is directly opposite the Sun. If that moment were exactly at the time of sunset for your location, you would see the full moon rise in the east just as the Sun was setting in the west. This rarely happens, however. In almost every month, what usually occurs is a discrepancy in the time at which the Moon is full and the time it rises at your location, even though the Moon appears fully illuminated.

The consequence is that the fully lit Moon will rise from a few minutes to an hour before or after the local time of sunset. The Moon, however, will appear full for at least a day before and after the exact moment when it is astronomically "full"—and sometimes for two or three days—because its spherical shape does not immediately begin to show the shadow lingering on the left side just before full moon or the shadow on the right side just after full moon. The period of apparent fullness is related to how fast the Moon is traveling in its orbit and the angle of the Moon's orbit compared to the Sun. Both of these vary from month to month.

No matter what the astronomical time of the event, the Moon will appear full on the calendar day when the hour of moon rise is closest to the astronomical time of the phase. In general, if the marked time of moon rise is 9:00 A.M. or later, you will see the fullest Moon on the evening of that day; if the marked

In some months, the Moon will seem to be full a day before or after the actual event. In general, observers can spot the difference between an almost full moon and one that is completely full be looking for a slightly irregular crescent-shaped zone along the right side, where the shadow of the waxing gibbous phase is gradually shrinking.

Waxing gibbous moon
2 days before full moon

time is earlier than 9:00 A.M., you will see the fullest Moon on the previous evening. But the difference from one day to the next may be negligible and barely notable—if at all—to the eye.

No matter where you are on the globe, the time listed as Universal Time (UT)—also known as Greenwich Mean Time (GMT)—for any phase of the Moon is the exact time when that phase occurs. The time of full moons and new moons are usually listed in calendars and almanacs in this time format. This is the official local time at the Greenwich Meridian (0 degrees longitude, near London, England) that is used as a worldwide standard by astronomers.

There are no variations in latitude or longitude that affect the time when phases occur, but if you are not in the UT zone, you do have to convert these times to a local time zone to compensate for the differences in how civil time is used. In order to convert to local time, you add or subtract the number of whole hours determined by how many time zones are between you and Greenwich, England. (See chart on page 58.)

A recurring factor confuses the date when moon phases are reported by local media sources, particularly television news. Most, but not all, newspapers and radio and television stations use phase times that have already been converted for their region; the times of new moon and full moon are corrected for local time. But some almanacs, international publications, and local media use Universal Time when mentioning a particular moon phase, eclipse, or "blue moon" (see page 125), without acknowledgment, or even knowing the difference. Although other errors may crop up in reporting, most of the time mistakes made about moon phases come from a confusion about time zones.

The time zone mistake most often happens when a new or full moon occurs in the early hours of the morning in Universal Time—between midnight and 5 or 6 AM in the morning in London. Translated into local time on the East Coast or farther west, there is a gap of seven or more hours, with time zones in the U.S. pushing back the time of an event to the previous day. A full moon that occurs at 3:15 AM UT on November 7, for example, would occur at 10:15 PM Eastern Standard Time the day before, November 6.

Traditionally, full moons in fall months have been considered the biggest, brightest of the year. The harvest moon was so named because it provided addition hours of light during which farmers were able to complete their annual harvest. The harvest moon, not coincidentally, is linked to the month of September, a month when many crops have traditionally been harvested, and also when a major astronomical event occurs every year in the northern hemisphere— the fall equinox, also referred to as the autumnal equinox.

This event marks the point in the Earth's orbit around the Sun when the ecliptic—the apparent path of the Sun's movement across the sky—is halfway between its annual movement north and south. A few days after the Fall Equinox (which typically occurs on or about September 23) the length of day and night are almost equal, hence the name. The same situation occurs a few days before the Spring Equinox (which falls on or near March 19). For both Fall and Spring Equinoxes, the date when daytime and nighttime are almost equal is affected by an observer's latitude. The lag in days is due to atmospheric distortion on Earth and the width of the Sun — these effects combine to alter when and how sunlight reaches the Earth's surface.

Also at this time, the full moon rises at its shallowest angle to the horizon; the shallow angle makes it loom larger and longer near the horizon as it rises, yielding more apparent light. In reality, the greatest illumination from the full moon—or any other moon—is when it is overhead and closest to an observer.

At other times of the year, the Moon rises about one hour later from one day to the next. But close to the autumnal equinox, the angle of the Moon's path cuts this daily lag down considerably. From August through October—and peaking in the crucial month of September—the Moon may rise only about 15 minutes later each day, compounding its already powerful image. When the time of a full moon occurs very close to the equinox, the effect is magnified even further, and the greatest magnification occurs if the full moon, equinox, and perigee coincide.

LUNAR TILT

The Moon appears to be tilting, or changing position, as it crosses the sky. This visual effect is most apparent when the Moon is waxing and the lighted portion of the surface appears to turn from pointing to the west (right) to almost straight down as the Moon sets. The Moon is not actually turning, however, but is following a curved path across the sky, the path defining its orbit. Pictures or illustrations of the lunar phases usually show the Moon when it is at its zenith (highest point above an observer) and therefore oriented straight "up and down" relative to the earth.

The illustration below shows the actual visual appearance of the Moon as it rises and sets on the same night. Variations in this tilt are affected by the latitude of an observer. The farther north, the greater the affect. Below the equator, the Moon is still affected by this characteristic of its orbit, but because observers are seeing it "upside down" compared to the northern hemisphere, its image is also reversed from top to bottom.

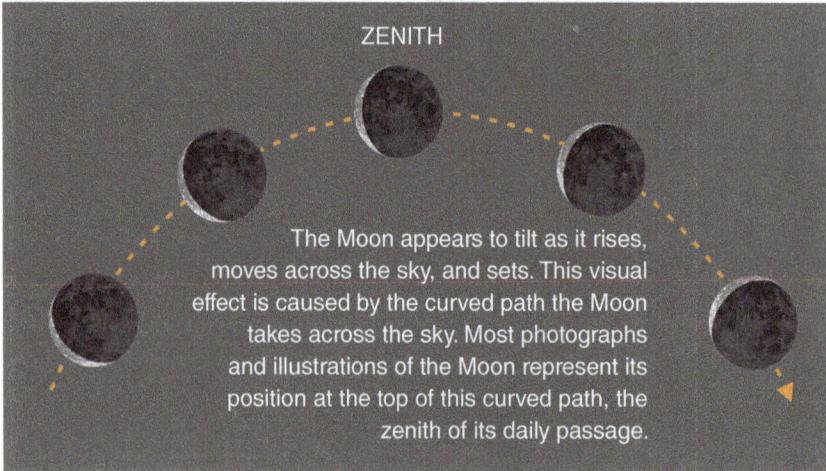

ZENITH

The Moon appears to tilt as it rises, moves across the sky, and sets. This visual effect is caused by the curved path the Moon takes across the sky. Most photographs and illustrations of the Moon represent its position at the top of this curved path, the zenith of its daily passage.

OCCULTATIONS

The sky is full of objects visible to observers on Earth. Planets and stars are visible as points of light to the naked eye. The planets appear to move at different rates than the stars as they are much closer to the Earth; stars are not actually moving at all, their apparent movement is due to the movement of the Earth. Both stars and planets, however, are not moving at the same rate as the Moon and a star or a planet will occasionally be obscured by the Moon when their paths overlap in the sky. This phenomenon is called an occultation.

Occultations occur whenever one body crosses in front of another body. Occultations caused by the Moon are the most obvious as the Moon's image is the largest in the sky. The most observable occultations to watch are those that occur when the Moon is less than full because there is less light, heightening the visual effect of the event.

Observers on Earth see an occultation as a point of light approaching the Moon from the left. The stars and planets appear to move across the sky faster than the Moon, accounting for this approach. Occultations can last for a few minutes or up to an hour, depending on the where they intersect the Moon, that is, whether the object appears to intersect the Moon at the widest part or just "grazes" the edge. At the instant of occultation, the point of light will suddenly disappear, appearing again when it emerges on the other side of the Moon. Grazing occultations appear as a blinking or twinkling phenomena, with the mountains and ridges on the Moon's surface momentarily breaking up the light from the star. When the large planets overtake the Moon, they may take a few seconds or longer to be occulted because of their size.

The brightest stars are also the most visible under the widest range of viewing conditions, but only those stars that lie in the zone of the Moon's path—about 5 degrees on either side of the ecliptic—can be intercepted by the Moon. Major stars and star groups likely to be observed being occulted by the Moon are Aldebaran, Antares, Regulus, the Pleiades, and Spica.

Lunar occultations are predictable because of the known orbital characteristics of the Moon. However, the times of occultations are difficult to compute because of the varying path of the lunar orbit and the range of possible loca tions of observers. Only the position of the stars remains unchanged. A distance of a few hundred yards to a mile north or south on the Earth's surface can make the difference between seeing a grazing occultation or a "near miss." The observer's longitude — east or west — determines the time when an occultation may be observed; the further to the east, the later the event.

Amateur astronomers who are serious about observing lunar occultations can participate in ongoing projects that track these events. Information is available at: The International Occultation Timing Association

www.lunar-occultations.com

During a grazing occultation, a star will appear to "blink" on and off as it moves behinds peaks and valleys along the Moon's horizon.

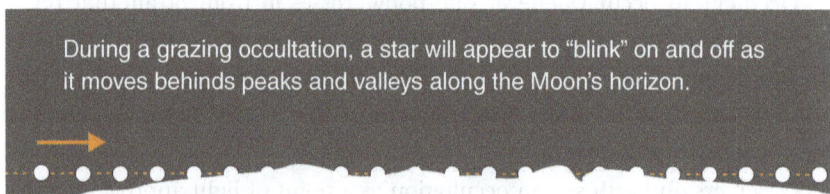

TWO TYPES OF GRAZING OCCULTATIONS

Grazing occultation on dark limb

Grazing occultation on bright limb

LIBRATIONS

The Moon always shows the same face to the Earth. Earth observers, however, get to see more than half of the surface of the Moon because of phenomena called librations. At any one time, the most that can be seen of the surface is slightly less than 50 percent because the spherical shape of the Moon hides the area closest to the perimeter. But librations allow the front surface of the Moon to be seen from slightly different angles at different times, producing an overall picture of the lunar surface that adds up, over time, to 59 percent of the total.

Librations are measured using longitudinal and latitudinal coordinates. Both are figured from a central point that is at a fixed geographical location on the lunar surface. This point is in the Sinus Medii, a small flat plain just below and to the right of Copernicus, a large rayed crater that is visible to the naked eye. Two meridians, one north and south (the Central Meridian) and one east and west (the Lunar Equator), cross at this point, which is also used to mark locations on maps of the lunar surface.

Different librations affect different zones of the Moon, with each type of libration contributing different degrees of added surface area. Depending on the orbital characteristics, the effect may vary from day to day and month to month. Sometimes, librations also overlap, creating an even stronger effect. The maximum added area that can be seen from a combination of librations is a little more than 10 degrees.

LIBRATION IN LONGITUDE

This libration is produced by the elliptical orbit of the Moon. If the orbit were a circle, the Moon's face would always point directly at the Earth and its orbital speed would remain the same. However, due to the nature of an elliptical orbit, the speed of the Moon changes depending on which part of the orbit it is in. This is also true of any object moving in an elliptical orbit. When moving from its fastest point (closest to Earth) to its slowest point (farthest from Earth), the Moon's speed is slowing down, but its rotation speed remains the same. For a period of time, the face of the Moon is therefore not pointed directly at the Earth. This "lag" effect allows observers to see an extra slice of surface, in effect,

63

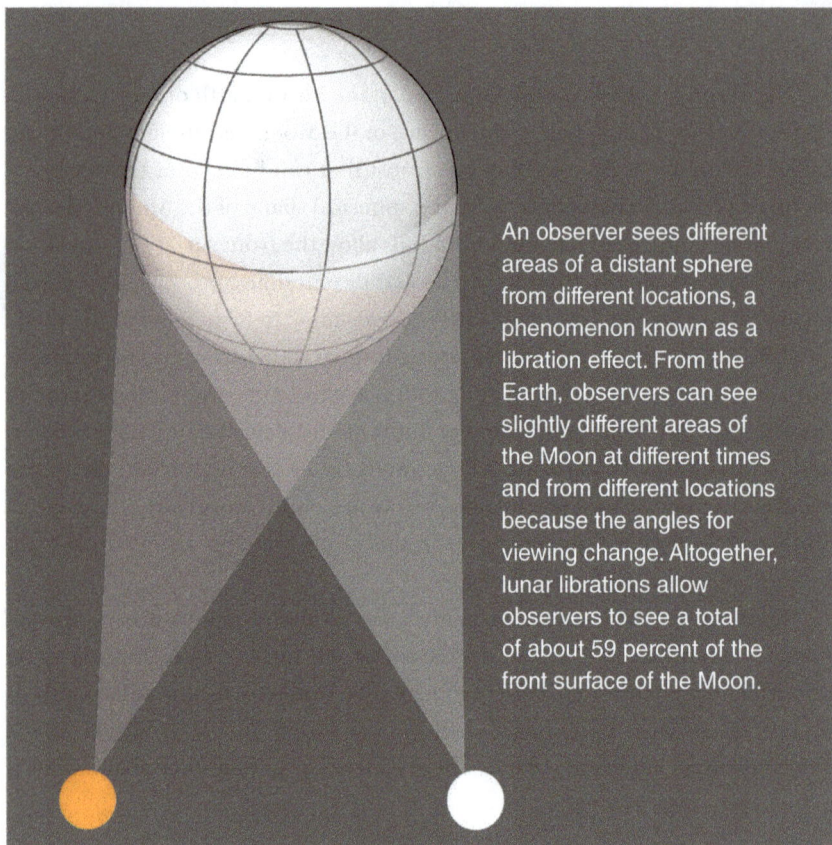

An observer sees different areas of a distant sphere from different locations, a phenomenon known as a libration effect. From the Earth, observers can see slightly different areas of the Moon at different times and from different locations because the angles for viewing change. Altogether, lunar librations allow observers to see a total of about 59 percent of the front surface of the Moon.

to "peek" around the edge of the Moon. When the Moon is at 90 degrees in its revolution (one-fourth of the way around), it is 97 degrees through its rotation. This libration is called longitudinal because the extra surface areas are exposed along lines of longitude (perpendicular to the equator). The total extra surface that can be seen with this libration is 8 degrees (7° 57', to be exact).

LIBRATION IN LATITUDE

The plane of the Moon's orbit is tilted about 5 degrees away from the plane of the Earth's orbit around the Sun. For half of a lunar cycle, the Moon is below the ecliptic, and for the other half the cycle it is above the ecliptic. Each of

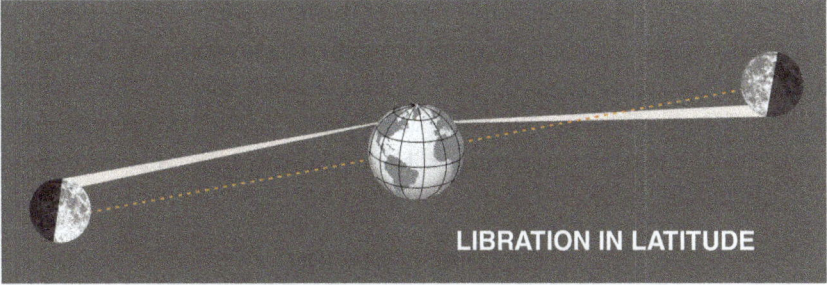

LIBRATION IN LATITUDE

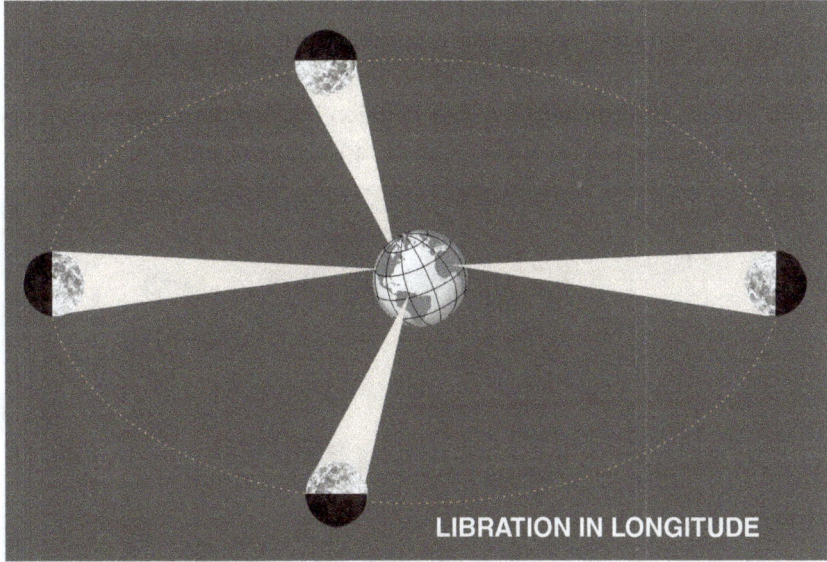

LIBRATION IN LONGITUDE

DIURNAL LIBRATION

these half cycles exposes an extra "slice" of the lunar surface at the top of its northern hemisphere or the bottom of its southern hemisphere, in effect allowing a "peek" above or below the normal limit of visible surface. This libration is called latitudinal because the extra surface areas that are exposed are great circles that are parallel to the equator. The total extra surface that can be seen with this libration is about 7 degrees (6° 51').

DIURNAL LIBRATION

Observers can "see over the top" of the Moon when it is rising, and "under the bottom" when it is setting. This is possible because the radius of the Earth adds an extra 4,000 miles of height advantage for looking "over" or "under" the Moon when it is on the horizon. This libration is called diurnal because it occurs every day, but only accounts for an extra 1 degree of visible surface.

Every lunar month, the libration in longitude alternately exposes more of the Moon's surface to the east and west. The maximum amount of additional surface that can be seen from this effect is a little less than 8 degrees.

Libration in latitude alternately exposes more of the Moon's surface to the north and south. A maximum of about 7 degrees of additional surface is seen from this effect.

TIDES

The Moon is the nearest celestial neighbor to Earth and exerts a constant influence through gravitational attraction. This force is only one ten-millionth of the gravitational force of the Earth itself, but combined with other forces—including the centripetal force created by the spin of the Earth—it is one factor that produces tides on the world's bodies of waters. There are also tides created in the atmosphere and, to a much lesser degree, in the Earth itself.

Lunar gravity does not work alone in producing tides. Tides are influenced by the centrifugal force created from the Earth's revolution around its barycenter (see page 1) and the gravitational attraction of the Sun. Gravitational attraction varies inversely with the square of the distance between two bodies. The Sun's mass, for example, is 27 million times larger than that of the Moon, but it is 390 times farther away from the Earth than the Moon. The result is that the Sun's gravitational force on the Earth is only 46 percent as much as the Moon's, making the Moon the most important factor for tides.

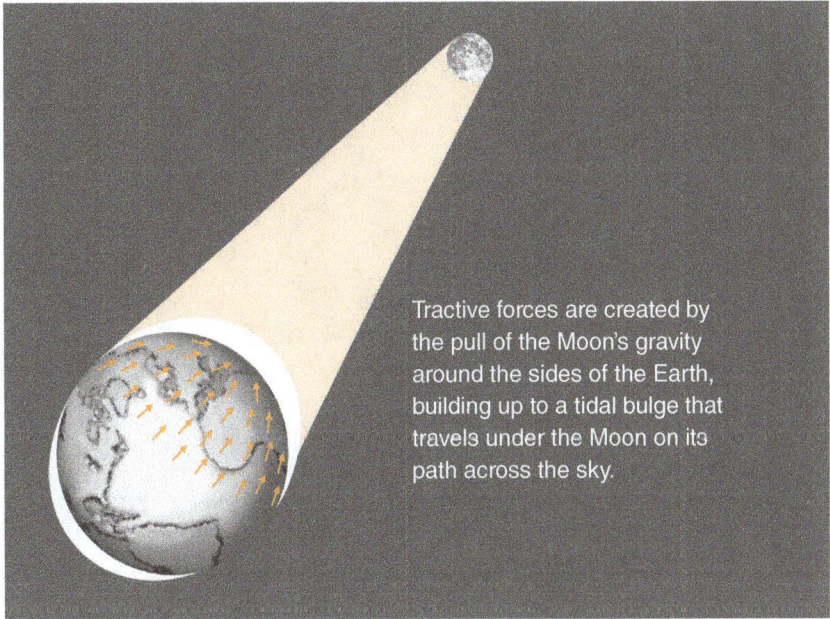

Tractive forces are created by the pull of the Moon's gravity around the sides of the Earth, building up to a tidal bulge that travels under the Moon on its path across the sky.

Tides, however, are not created by the direct pull of the Moon's gravity. The gravitational force of the Moon is tugging upward on the water while the gravitational force of the Earth, which is far stronger is pulling down at the same time. Instead, water rises in tides because of a net balance of forces—the Earth pulling in and the Moon pulling out—which averages more in favor of the Moon. It does not happen in a perpendicular direction, however, but shows up where the external influence has a greater effect, from the side.

This type of gravitational attraction is known as tractive force. In a simplistic sense, it is the same phenomenon experienced if a person attempted to pick up a heavy object from above. In order to lift the object, it requires enough strength to completely overcome the weight, or mass, of the object.

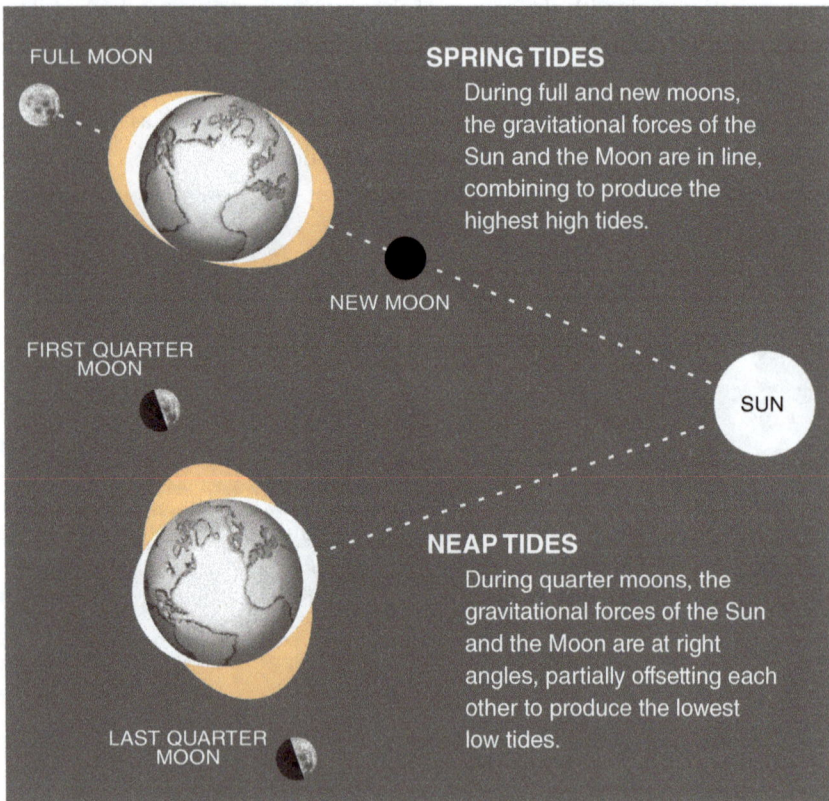

FULL MOON

SPRING TIDES
During full and new moons, the gravitational forces of the Sun and the Moon are in line, combining to produce the highest high tides.

NEW MOON

FIRST QUARTER MOON

SUN

NEAP TIDES
During quarter moons, the gravitational forces of the Sun and the Moon are at right angles, partially offsetting each other to produce the lowest low tides.

LAST QUARTER MOON

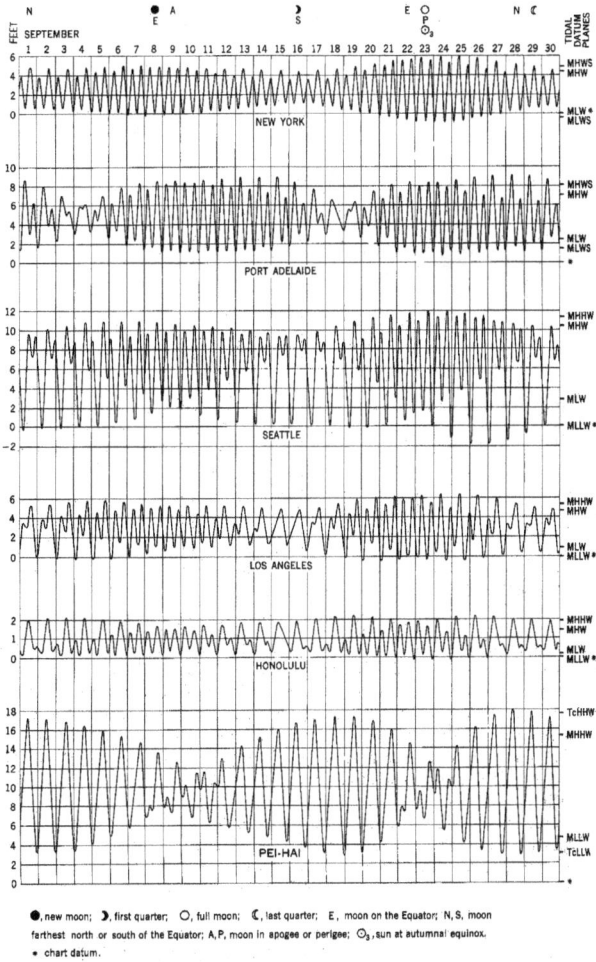

●, new moon; ☽, first quarter; ○, full moon; ☾, last quarter; E, moon on the Equator; N, S, moon farthest north or south of the Equator; A, P, moon in apogee or perigee; ☉₃, sun at autumnal equinox.
  ● chart datum.

Tide tables are a traditional published resource, offering local times of high and low tides a year or more in advance. In the modern era, this vital information is increasingly available online and through apps used on portable electronic devices.

But if a person pulls the same object from the side, sliding it over a surface, for example, it takes much less force to move it.

Tractive forces create a "piling" effect on the Earth's oceans, pulling water toward the Moon from around the circumference of the planet, adding up to the greatest effect—and the highest pile—closest to the spot that is underneath the Moon's position, its zenith. Because of the motion and relative positions of the Earth, the Sun, and that of the Moon, this pile, or high tide, is typically somewhat ahead or behind the actual zenith of the Moon. The time it takes for this high water to arrive before or after the Moon passes overhead is called the lunitidal interval.

At the same time that water has piled up on the Moon's side of the Earth, a second bulge has piled up on the opposite side of the planet. This second bulge is a result of water moving to create an equilibrium between the gravitational force of the Moon and the centrifugal force created by the movement of the Earth. Opposite from the Moon, the net effect on the oceans is to pull away from the Earth. Because the Moon's orbit around the Earth is completed about once every 25 hours, each of the two tidal peaks—as well as the two tidal troughs—occur halfway through that period, about 12½ hours apart.

PERIHELION TIMELINE

Year	Date	Time
2016	Jan. 2	10:48 PM UT
2017	Jan. 4	2:17 PM
2018	Jan. 3	5:34 AM
2019	Jan. 3	5:19 AM
2020	Jan. 5	7:47 AM
2021	Jan. 2	1:51 PM
2022	Jan. 4	6:55 AM
2023	Jan. 4	4:18 PM
2024	Jan. 3	12:39 AM
2025	Jan. 4	1:29 PM
2026	Jan. 3	5:16 PM
2027	Jan. 3	2:33 AM
2028	Jan. 5	12:29 PM
2029	Jan. 2	6:14 PM
2030	Jan. 3	10:13 AM
2031	Jan. 4	8:49 PM
2032	Jan. 3	5:12 AM
2033	Jan. 4	11:52 AM
2034	Jan. 4	4:48 AM
2035	Jan. 3	2:18 PM

The Sun's gravity also produces daily tides. But because the force is smaller than that of the Moon, the effect produces smaller tidal bulges. The Sun's daily cycles are also slightly shorter than the Moon's, with only 12 hours between solar tide peaks. The Sun, however, can have an additional effect on tides based on how close the Earth is to it during the annual orbit. The Earth's path around the Sun is an eclipse (although not as extreme as the Moon's elliptical path around the Earth) and from December through January, the Earth comes its closest to the Sun. The point of closest approach is called the perihelion and in this century, falls between January 1 and January 5 in the western hemisphere.

Shortening the distance between two object magnifies the gravitational pull; at perihelion, the Sun's gravitational force is strongest. High tides that happen near perihelion are called "king tides," because they are typically the highest high tides of the year. On rare occasions, the day of a perihelion may coincide with a new or full moon, making for a potential extra-high king tide — and as part of this cycle, an extra-low low tide as well.

Tides are made more complicated by the effects of local geography. The speed and height of tides are affected by water depth, wind, and the obstructions on shorelines and below water. In the open ocean, the total difference between high and low tides is about one foot. Along some coasts, the difference can be more than twenty feet.

When the Sun, Moon, and Earth are in a line, the gravitational effect of the Sun adds to that of the Moon, creating the greatest tidal effect. This kind of alignment happens twice every lunar month because it is the same alignment that produces a new moon and a full moon. If the Sun and Moon are on the same side of the Earth (new moon) or on opposite sides (full moon), a spring tide is the result. Spring tides create the highest high waters and lowest low waters every month. When the Sun and Moon are at right angles (first and last quarter moons), a neap tide is the result, when the difference between high and low waters is at a minimum. When the Moon is closest to the Earth every month, its perigee, gravitational forces are also greater, but the largest effect on tides only comes when this orbital event coincides with a full or new moon, producing what is known as a perigean spring tide. For more on perigean spring tides, see the following section.

SUPER MOONS

In recent years, the media has increasingly mentioned the pending arrival of a "super moon," a full or new moon that occurs close to the closest point in the Moon's monthly orbit, its perigee. Perigean spring tides are part of super moon events, but they are nothing more than the traditional perigean spring tides. At its closest point in orbit, the Moon's perigee also yields a Moon at its largest diameter, another point of emphasis when these events are heralded, with the anticipation of viewing a larger-than-normal lunar disc.

Despite reports in the media in advance of super moons, the extreme tide potential does not usually play out, because tides are affected by a number of factors that often work to reduce effects on local shorelines. Weather is perhaps the most critical factor; the right kind of storm conditions can exacerbate the extra-high times or offset their effects, depending on wind velocity and direction and other factors.

The biggest issue with super moons is a mistaken application of the concept. According to some weather forecasts, almanacs, and astrologers, a super moon occurs any day there is a new moon and a perigee—the term comes from astrologers, not astronomers. In reality, the pairing of these events rarely happens close enough together to produce the compounded force needed to significantly affect tides. In order for the nearness of the Moon to have its greatest effect, the time of convergence of perigee and new or full moon must be within an hour or two, something that happens much less often than just a calendar day on which the two events occur.

This nearness of this conjunction is important because of the physics controlling gravitational attraction—the measured effect between two bodies is not linear, but exponential. That is, the force picks up or drops off considerably as the two bodies move closer together or farther apart. The same effect governs light.

Perigean spring tides are about as rare as blue moons, occurring about once every few years, on average. But because of the particulars of the lunar cycles, a perigean spring tide will be followed by others, 6½ and 7½ lunations later. On January 4, 1912, a lunar perigee (1:35 PM UT) and a full moon (1:29 PM UT) coincided and the two events were within six minutes of each other. The perigee on that date was also the most extreme in more than 1,400 years, a distance of 220,953 miles (356,375 km). Compounding the tidal effect from this proximity, the Earth was also at its closest point of the year to the Sun (1:35 PM UT), a point known as the perihelion, adding an additional force from the Sun's gravity. According to an estimate generated by *Sky & Telescope* magazine, this effect generated a tidal force that was 74 percent stronger than the benchmark set by the Moon at an average (mean) separation from the Earth.

Although the very rare 1912 tidal event did not result in historic flooding, some modern astronomical sleuths believe there could be a link between the occurrence on January 12 and something that is in the history books, the sinking of the *Titanic* on April 14 of the same year, after striking an iceberg. This theory suggests that the extreme perigean spring tide was a factor involved in what was noted at the time as a season with an unusually large number of drifting icebergs. However, the time delay between the calving of icebergs and the distance they had to travel to impact the *Titanic* tempers this idea. If such a super moon is worth worrying about, there is plenty of time for anticipation before such an event happens again. The next such extreme perigean spring tide is not due until 2257.

As for size, there are also other factors at play when the Moon comes closest to Earth, diminishing most of the visual effect. For one, the eccentrically shaped orbit of the Moon is itself rotating—the high and low points of the orbit move over time, taking a little more than 18 years to repeat the same pattern. This means that the low point in the orbit is not the same from month to month and there is a single extreme low point during a period of a year. The difference in Moon size from its closest extreme to its farthest yields a maximum difference in size of about 10 percent—barely enough to notice—but this maximum difference does not apply from month to month. A super moon is most often only a few percent or more larger in size than a non-super moon.

VIEWING THE MOON

The Moon is by far the largest single object in the night sky (not counting the Milky Way) and some of its major surface features can be identified with the naked eye. Binoculars, telescopes, and spotting scopes, however, greatly enhance the viewing experience, and because of the Moon's size and closeness to observers on Earth, there are plenty of options for accessible, affordable equipment.

BINOCULARS

Even the smallest, low power binoculars can provide dramatic, detailed views of the Moon. Field glasses, also known as opera glasses (or technically, Galilean binoculars) lack the prism found in "true" binoculars, and yield lower magnification. True binoculars use prisms to correct images from the two barrels to an upright position; the images are combined by the observer's eyes. The two basic types of true binoculars are roof prisms and Porro prisms; Porro prisms are generally considered superior for astronomical viewing. The specific capabilities of binoluclars are noted with two key numbers—in a form such as 7x35, 7x50, 10x50, etc.—with the first number representing the magnifcation and the second the aperture (the diameter in millimeters of the front lens).

Binoculars with the best performance at night need the largest aperture because the wider the front lens, the more light can be captured; resolution is a function of aperture. Magnification is of benefit if an observer wants to see the maximum amount of detail on the lunar surface.

In general, the higher the magnification, the narrower the field of view. There are no barriers to using a single pair of binoculars for spectator sports, bird watching, and viewing the Moon, but when viewing celestial objects, a wider field of view is usually preferable. Except for the highest power binoculars, most units will have a field of view of 5° to 8°.

Wth higher magnification comes an unwanted side effect: vibration, or shake. As a rule of thumb with astronomy, binoculars that are more powerful than 10x50 are hard to hold steady. Just as with telescopes and cameras, a tripod, monopod, or other solid support provides a stablizing effect that improves viewing and photography.

Technology does provides an option. Image stabilization (IS) is available with some binocular brands, though it does add to the price. Image-stablized binoculars combine traditional optics with digital sensors and advanced technology to greatly reduce the amount of shaking for hand-held viewing; the technology also improves on tripod-mounted use. Prices for image-stablized models can be several hundred dollars more than traditional models—high-end models with the best attributes for celestial viewing can cost more than $1,000.

Waxing crescent moon photographed with a DSLR and 200mm telephoto lens shooting through 20x80 binoculars.

IS binoculars are heavier than the traditional type and require batteries.

SPOTTING SCOPES / MONOCULARS

A spotting scope is a small telescope designed for terrestrial viewing. The main characteristic of such a unit is an angled eyepiece for more convenience when looking at birds or other wildlife in near-horizontal positions; most telescopes, on the other hand, are equipped with eyepieces set at a 90 degree angle. Viewing angle notwithstanding, a spotting scope can be practical for viewing celestial objects, including the Moon, with the advantage of portability.

Monoculars are smaller than spotting scopes and provide maximum portability. The smallest are about the size of a pack of cigarettes, but as with binoculars, viewing results are directly related to the size. Power ratings are stated with the same formula: 8x32 means a magnifying power of 8 and a front objective that is 32mm in diameter. Monoculars are not a preferred system for viewing celestial objects in general—at least compared to the other optic options—but because the Moon is so large compared to the planets, they can provide credible results.

TELESCOPES

For celestial observations, telescopes are generally rated by their ability to capture light—the greater the light-capturing, the more visual information the telescope can reveal. But the Moon is an exception to this rule of thumb, because it is the brightest object in the night sky and any telescope will produce enhanced viewing. As a general rule, Moon observers will benefit most from the resolving power (resolution) of the scope, not the light-gathering capability.

Of the major types of telescope technology, the refractor design is the most traditional and also the one most amateur and professional astronomers prefer for lunar viewing, although any of major designs work well. The most appropriate consideration when the Moon is the target object is not the size of the unit (typically the aperture size), but the quality of manufacture. Even the smallest telescope yields acceptable results if the mechanical construction and optics are well-made; as with most tools, you get what you pay for.

The smallest refractors typically have 3-to-4 inch (70-100mm) apertures and will provide significant detail on the Moon's surface; they are also light enough to be portable. Telescopes require tripods and specialized mounts to provide good results; many entry-level products comes as a bundled unit with telescope, tripod, and mount included.

MOUNTS

When magnified, the Moon has one characteristic—movement—that makes a specialized telescope mount essential. Even with a telescopic lens on a camera, the image of the Moon drifts out of frame quickly enough to require constant re-aiming, although short exposures are the rule for this bright object. And in any case, magnification not only enlarges the viewing object, it magnifies vibrations. A rule of thumb: the greater the magnification, the greater the shakes and the faster the observed object will move across the field of view. Although manual re-aiming can effectively deal with the Moon's movement for many viewing experiences, a magnification of 100x or higher will typically generate enough movement to greatly affect the viewing experience.

The basic altazimuth mount (short for "altitude" and "azimuth," the two perpendicular axes) is essentially a swiveling yoke. A manual knob allows the

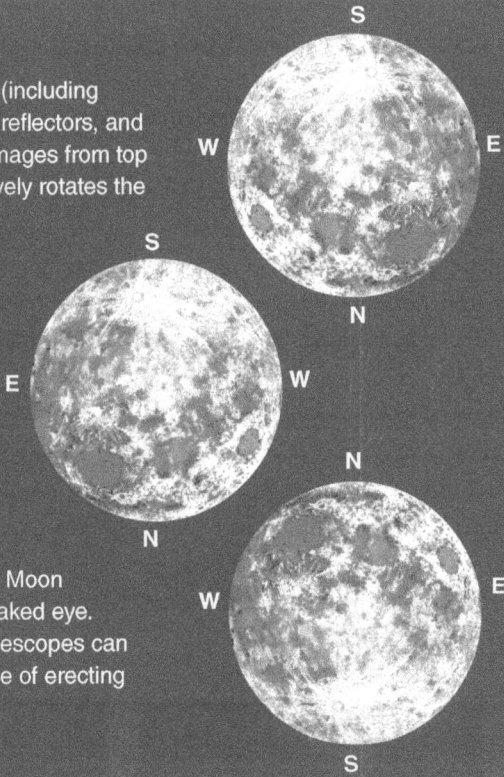

Almost all telescopes (including refractors, Newtonian reflectors, and catadioptrics) invert images from top to bottom. This effectively rotates the image 180 degrees.

Some telescopes (refractors and catadioptrics) also reverse images from left to right.

The orientation of the Moon as it appears to the naked eye. Inverted images in telescopes can be altered with the use of erecting prisms or eyepieces.

user to keep up with the travel of a target object along its path through the sky; many such mounts are motorized and once aligned, automatically offset the movement of the Earth and the Moon. Equatorial mounts are a different design that also automatically track celestial objects.

The latest computerized mounts compensate for the movement of the Earth by locking on to key stars to keep target objects, such as the Moon, within the telescopes viewing frame; some of the newest equatorial designs add GPS input to further support alignment, as well as simplify the setup process for users.

Mounts add cost to the purchase of a telescope, although there are many telescope models that come bundled with both tripod and mount. Prices range from a few hundred to thousands of dollars.

Viewers are likely to have different goals when gazing at the Moon. Some are interested in examining details on the surface while others want to see the entire disk, as during a lunar eclipse; another common objective is the full disc of the Moon as well as nearby planets, stars, or star groupings. Field of view is thus an important consideration, although this can be modified on the fly with different eyepieces in the same telescope.

When looking through a telescope or binoculars) at night, the pupil of the human eye becomes smaller as the Moon becomes brighter. The eyepiece of a telescope should be sized so that the viewable image is the same size or smaller than the viewer's pupil. This ratio will yield the maximum resolution from the telescope.

In the terminology of telescopes, the "entrance pupil" is the aperture where light enters the optics; the "exit pupil" is the lighted disc of an image that floats just behind the eyepiece, where a viewer's eye would be. If the exit pupil is larger than the viewer's pupil, the effect is to reduce the aperture of the telescope, resulting in a loss of detail. In humans, pupil size varies from person to person and flucuates with age. In general, the older the person, the smaller the pupils; the typical range is 5 to 7mm.

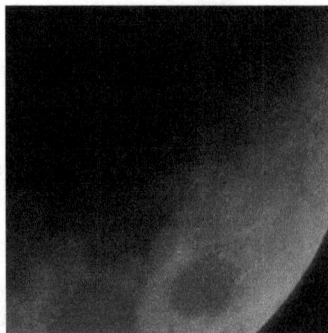

In many cases, a spotting scope or small telescope can reveal features on the Moon's surface with more clarity and detail than larger, more powerful instruments. The photographs above were taken with a DSLR and 20-70mm lens shooting through an 82mm spotting scope — the photo on the right has been enlarged with software.

PHOTOGRAPHING THE MOON

The Moon is the biggest and brightest object in the night sky, and as such, attracts the interest of many photographers. But even when the Moon is near the horizon and seems to be quite large, it is difficult to capture with a camera what is seen with the naked eye. In fact, the most common problem photographers have with this celestial object is its size. The naked eye sees the Moon roughly equivalent to a 50mm lens, but with a standard 50mm lens, the final image will be less than 1/50 inch in diameter (less than half a millimeter).

Longer lenses will give better results and telephoto lenses are the next best thing to a telescope. With camera lenses, the rule of thumb is that anything longer than 400mm requires a tripod to reduce vibrations; modern lenses with image stabilization manage this problem somewhat, but tripods are still recommended.

To determine the final size of the Moon in photographs, use this formula:

Final image size = focal length ÷ 109

For example:	Lens Size	Image Size
	100mm	0.92mm
	200mm	1.83mm
	300mm	2.75mm

As well as longer lenses, converters multiply the power; a 2x teleconverter will double the focal length of a 200mm lens to 400mm. However, converters also double the f-ratio, requiring longer exposures. For example, a 100mm lens at f/8 becomes 200mm at f/16 using a 2x teleconverter. But because the Moon is so bright—at least when it is full—this does not represent much of a problem.

Exposure time for lunar photos is important because the Moon is in motion across the sky. Telescopes with motorized mounts—and cameras fixed to these

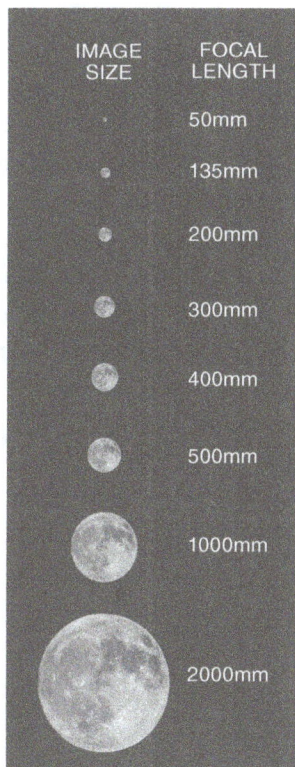

IMAGE SIZE	FOCAL LENGTH
	50mm
	135mm
	200mm
	300mm
	400mm
	500mm
	1000mm
	2000mm

devices—compensate for this movement. Without such a drive, the maximum exposure time should be set as fast as possible to eliminate blurring. A rule-of-thumb formula: Exposure (in seconds) = 250 ÷ f-stop x focal length in mm

The brightness of the Moon varies according to its phase, further complicating exposure times. Another formula to deal with this is based on the Moon's phase: Exposure (in seconds) = $f\text{-stop}^2$ ÷ (A x B)

Variables: A = ISO number, B represents brightness (10 for thin crescent, 20 for wider crescent, 40 for quarter moon, 80 for gibbous moon, 200 for full moon)

Atmospheric conditions and the relative illumination of the Moon at different distances in its orbit make formulas a less than perfect strategy; the varying quality of different lenses and cameras also have a big impact. Experimentation and bracketing (a series of shots with different exposure times) are a prudent option and one of the great advantages of digital cameras.

Shooting from the surface of the Earth through its atmosphere alters the depiction of colors, especially on the edges of object, and the quality of optics is also a contributor. Chromatic aberration (CA) is also a recurring issue; photographs taken of celestial objects, including the Moon, can be distorted by the equipment itself. High quality optics can keep CA under control, but many lunar photographs benefit from a little tweaking with graphics software to deal with unwanted fringing. Photographs of the Moon are also often taken as black and white images or converted to B&W with software later to remove such unwanted visual effects.

CAMERAS

The simplest method for lunar photography is by aiming a camera through a telescope or binoculars. This can work with either a point-and-shoot digital camera, a DSLR (digital single-lens reflex), or a video camera. This method is called digifocusing or the afocal method. Using the viewfinder or the LCD screen, aim the camera at the eyepiece, keeping them close together but without contact. Trial and error are part of this approach — keep trying until the results are satisfactory. The telescope or binoculars should be mounted on a tripod, and an adaptor to hold the camera to the eyepiece is recommended. It may take several shots to get the image lined up correctly in the camera.

In the digital era, the general rule is that more pixels are better, and DSLRs with full-frame sensors are preferable, but not required. The most important factor is not the camera, but the lens. The size (diameter) of the objective lens and its quality are the key deciding factors, and as with most optics, the bigger and better, the costlier. Telephoto lenses that are longer than 400mm, even with image stabilization, perform much better with tripods, and DSLRs will outperform pocket cameras, as much for their ability to capture more pixels as their manual adjustments, which allow more effective control. If a pocket camera is used, optical zooms are the preference; the digital zoom function is unlikely to produce quality results.

Full moon photographed with an iPhone 5 and an 82mm spotting scope.

The best strategy with lunar photography is to take lots of photos, adjusting the settings to produce a selection that is most likely to yield optimal results.

ADAPTERS

Handheld shooting through the eyepiece of a telescope or binoculars can be improved if the camera and lens are firmly anchored to a stationary object such as a tripod. Telescope manufacturers provide adapters that also allow many kinds of cameras to attach directly to the a telescope eyepiece. These usually feature threaded bases or insets and the thread specifications typically follow several universal standards, but both the eyepiece and the camera lens have to match for this mating to work. There are similar adapters for binoculars.

VIDEO

In the modern era of digital photography, digital video greatly expands possibilities for capturing images of the Moon. Although dedicated equipment is available for this purpose, standard camcorders and the video functions on many consumer DSLR cameras can provide satisfying results. Video footage of lunar eclipses is one application, with "fast forward" of the final output providing results similar to traditional time lapse exposures from still cameras.

Another application for video that is increasingly used by amateur astronomers is video footage of celestial objects—including the Moon—that is processed with specialized software that "stacks" multiple images to produce a final still frame that has enhanced details.

Dedicated video cameras for telescopes are typically mounted with device-specific hardware, including specialized cameras and mountings. But successful results are also achievable with standard cameras, even when they are hand-held to capture images through an eyepiece. Guidelines and "how to" advice is readily available on Web sites and astronomy blogs. Lunar photography benefits from trial and error.

PHOTOGRAPHING ECLIPSES

Location is a major factor in planning for an eclipse. Eclipse maps are a useful source in this endeavor and are readily available online, sometimes several years in advance. However, most eclipse maps provide only a general sense of geographic zones involved and for solar eclipses in particular, detailed local maps of the eclipse path are critical for planning. These can often be found through astronomy groups, which typically post maps online months in advance. *Sky & Telescope* Magazine is also a key source.

Location selection can be facilitated by using Google or other online mapping resources that provide a "street view" option, allowing you to spot potential landscape features that are desirable, as well as those that might not be. Flickr and other popular Web-based photo-sharing sites are also useful tools for scouting, with users searching for geographic locations that represent eclipse viewing targets.

Advance plans can also be enhanced with the use of a computer app that provide accurate sight lines and compass headings. These programs—The Photographer's Ephemeris is one such resource—allow users to plug in coordinates for an anticipated viewing spot and see the rise and set points for the Sun and Moon, as well as the track of the ecliptic. Advance weather information, also found online, is equally important for preparation.

DIGITAL CAMERA RULES OF THUMB

- A tripod is always recommended.

- 200m telephoto lens or larger, prime or zoom.

- No filters. No flash.

- Remote trigger, tethered release cable or wireless.

- For mirror-based DSLRs, turn off or lock the mirror to reduce camera shake.

- Do not use automatic metering. Auto metering will typically cause overexposure. If available, the Spot Metering Mode can be used for shots of the full moon if the meter reading is centered directly on the center of the Moon.

- ISO: 100-200 (turn off Auto ISO). Faster ISO settings produce more noise, which may not be fixable with software.

- Aperture: f/8 to f/16.

- Shutter speed: 1/125 to 1/200.

- Use manual focus. Set the focus to infinity but test results but if the camera has a "live view" function, use this function plus magnification to fine tune the focus.

- If the Moon is too bright, adjust the shutter speed to a faster setting; if it is too dim, adjust the speed to a slower setting. To limit motion blur, shoot at 1/100 or faster. The longer the focal length of the lens, the faster the shutter speed.

- Auto white balance is often adequate for most cameras, but if a test shot indicates that Moon image has too much orange/red, change the white balance setting to Tungsten. If your camera has an auto setting for "Cloudy," this may also produce good results. The final color can be adjusted in a photo-processing program.

- For handheld shots, increase the ISO to 800-1000, decrease the aperture to f8 or f9, and use a faster shutter speed—1/1000 to 1/1500 second—to compensate for the lack of a tripod support.

When shooting lunar eclipses, the same general guidelines apply as for regular lunar photography, with extra care paid to exposure times. As the full moon passes into the Earth's shadow, its illumination drops considerably, requiring an adjustment to camera settings. As a rule of thumb: 1/250 for first contact with the umbra; 1/125 for partial umbra; and 5–15 seconds for full umbra.

Time-lapse images depict a sequence as the Moon or Sun enters and leaves the event provide a visual guide to the event. This concept is enhanced by not moving the camera, using a 50mm or 85mm lens that covers the full span of distance that the Moon or Sun traverses. The exposure does have to be changed during the movement, however, to compensate for the fluctuating illumination. Planning is more important with this goal in order to make sure the placement of the camera and the field of view with the designated lens will cover the intended results.

Solar eclipses follow a similar pattern, with only a single significant rule: always use a certified solar-rated filter. Cameras and human eyes can be seriously—and permanently—damaged by exposure to an unfiltered Sun. The right filters are those designated for specific lenses; jury-rigging lens coverings may permit damaging light leakage. The standard for solar viewing through telescopes or photography of eclipses is a front-mounted filter rated at neutral-density 5 (ND-5). Eclipse-specific filters are typically aluminum-coated Mylar, nickel-chromium coated glass, or hydrogen-alpha.

The apparent size of the Sun is about the same as the Moon, making the choice of lenses similar.

DIGITAL IMPROVEMENT

Photos of the Moon can be improved with software. Programs such as Photoshop or its equivalent provide basic digital tools that can greatly enhance images and in some cases, fix flaws that are generated by viewing conditions or equipment.

One of the most sought-after photos is of the Moon when it is close to the horizon; an image that includes both the lunar disc and elements of the landscape. But the full moon is just too bright to be properly exposed within the same shot as terrestrial features as they are framed in the dim evening light. Either the landscape will be too dark or the Moon will "wash out" as an overexposed blob.

The only reliable way to shoot a full moon when it is close to the horizon is with a double exposure, a process accomplished with software. One shot should follow the standard guidelines for a full moon, exposing for a full range of surface detail, shadow, and highlights; a second shot, without moving the camera on the tripod, exposes for the foreground. The two shots are combined using layers and blend modes to capture the best elements of each.

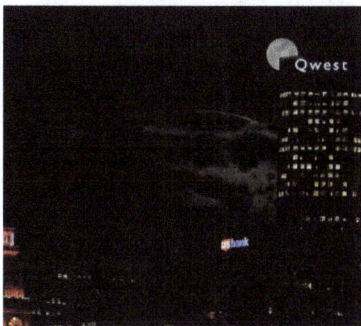

To display the best features of a full moon and a landscape—or cityscape, as above— a double exposure is required.

IMAGE PROCESSING

- **Enlargement** Avoid temptation and limit englarging to less than 150%. Final files should be 72 or 96dpi for posting online; 300dpi or more for prints. Resize without resampling to avoid degrading the original. Specialized enlargement software such as Genuine Fractals is recommended.

- **Contrast** Increase by 10-20% or use a "curves adjustment" tool to add medium contrast.

- **Sharpen** Increase the sharpness of the Moon itself with a dedicated sharpness tool or use an "unsharp mask" filter to add 125-150% coverage, but reduce the radius to 1.0 pixels or less. High pass filters also provide a controllable adjustment.

- **Dodge/Burn/Sponge** These tools can be used to selectively lighten or enhance specific areas, such as craters or valleys.

- **Desaturation** Use a desaturation tool to deal with chromatic abberations, such as color fringing that often appears around the circumference of the Moon itself, or the edges of specific surface features.

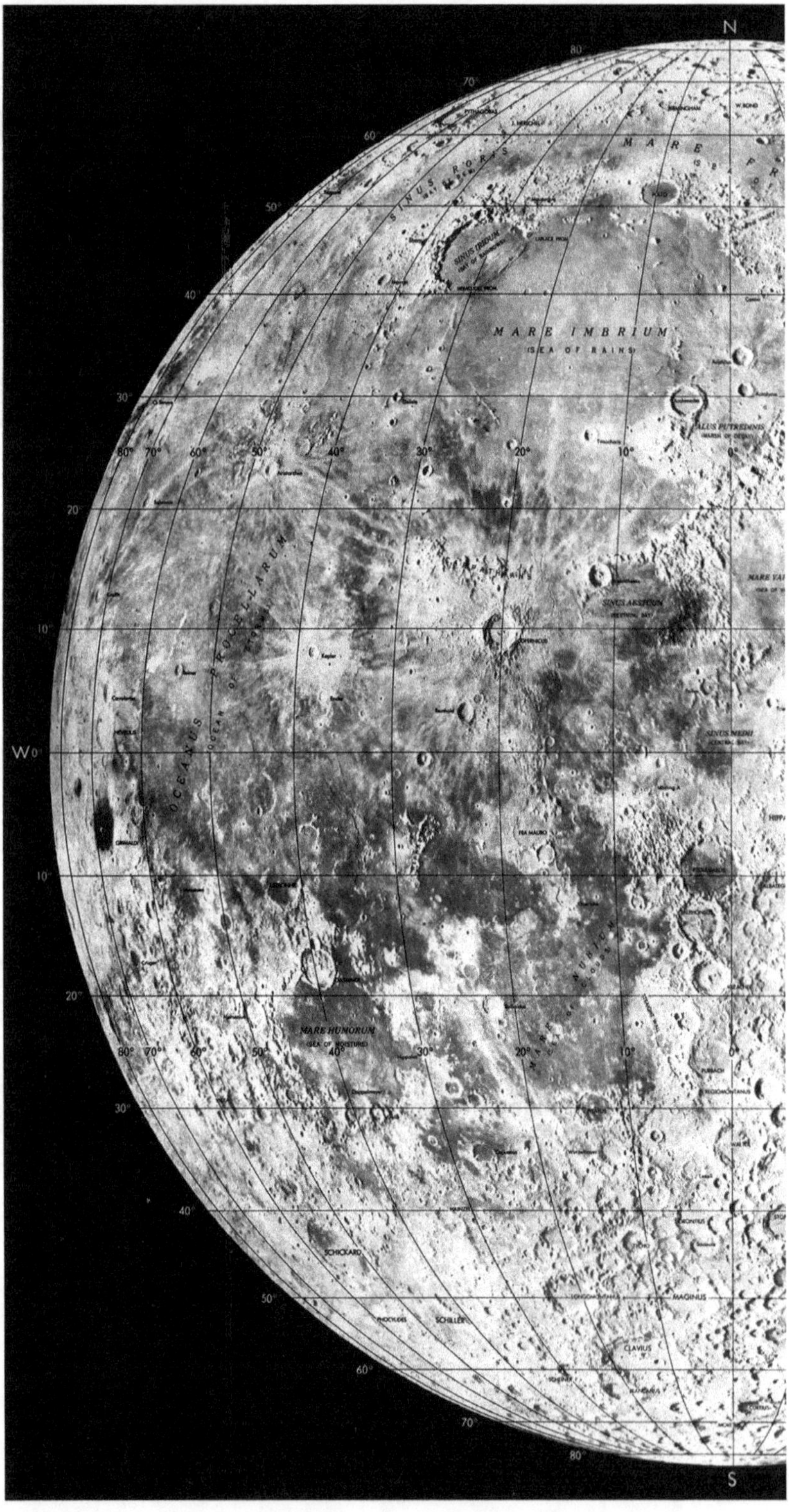
N
S
W 0°
MARE IMBRIUM
(SEA OF RAINS)
SINUS IRIDUM
OCEANUS PROCELLARUM
OCEAN OF STORMS
MARE HUMORUM
(SEA OF MOISTURE)
GRIMALDI
FRA MAURO
SCHICKARD
SCHILLER
MAGINUS
CLAVIUS

LUNAR GAZETTEER

* Archimedes (crater diameter: 50 miles) 30° North, 4° West
* Aristarchus (crater diameter: 25 miles) 24° North, 48° WestAristillus (ray crater; crater diameter: 35 miles; ray pattern diameter: 400 miles)34° North, 1° East

Aristoteles (crater diameter: 55 miles) 50° North, 18° East

Arzachel (crater diameter: 61 miles) 18° South, 2° West

Atlas (crater diameter: 54 miles) 47° North, 44° East

Bailly (crater diameter: 184 miles) 66° South, 65° East

Barocius (crater diameter: 54 miles) 45° South, 17° East

Berosus (crater diameter: 45 miles) 33° North, 70° East

Blancanus (crater diameter: 72 miles) 64° South, 21° West

Colombo (crater diameter: 46 miles) 15° South, 46° East

Condorcet (crater diameter: 49 miles) 12° North, 70° East

* Copernicus (ray crater; crater diameter: 57 miles; ray pattern diameter: 750 miles) 10° North, 20° West

Endymion (crater diameter: 77 miles) 55° North, 55° East

* Gassendi (crater diameter: 69 miles) 18° South, 40° West

Grimaldi (crater diameter: 127 miles) 6° South, 68° West

Gruemberger (crater diameter: 58 miles) 68° South, 10° West

* Gutenburg (crater diameter: 46 miles) 8° South, 41° East

Hipparchus (crater diameter: 95 miles) 6° South, 5° East

Hommel (crater diameter: 75 miles) 54° South, 33° East

Humboldt (crater diameter: 130 miles) 27° South, 81° East

* Kepler (ray crater; crater diameter: 20 miles; ray pattern diameter: 400 miles) 8° North, 38° West

Kircher (crater diameter: 48 miles) 67° South, 45° West

Lacus Somniorum 37° North, 35° East

Lacus Mortis 44° North, 27° East

Langrenus (ray crater; crater diameter: 82 miles; ray pattern diameter: 950 miles) 9° South, 61° East

Letronne (crater diameter: 73 miles) 10° South, 43° West

Longomontanus (crater diameter: 92 miles) 50° South, 21° West

Maginus (crater diameter: 116 miles) 50° South, 6° West

Mare Australe 50° South, 80° East

* Mare Crisium 18° North, 58° East
* Mare Fecunditatis 4° South, 51° East

When the Moon is not full, such as the first quarter moon
seen above, many of its features are more pronounced and
details are easier to see because of the shadows formed by
oblique lighting.

* Mare Frigoris 55° North, 0° East
  Mare Humboldtianum 55° North, 75° East
* Mare Humorum 23° South, 38° West
* Mare Imbrium 36° North, 16° West
  Mare Marginis 13° North, 87° East
* Mare Nectaris 14° South, 34° East
* Mare Nubium 19° South, 14° West
  Mare Orientale 19° South, 95° West
* Mare Serenitatis 30° North, 17° East
  Mare Smythii 3° South, 80° East
  Mare Spumans 1° North, 65° East
* Mare Tranquillitatis 9° North, 30° East
* Mare Vaporum 14° North, 5° East
  Maurolycus (crater diameter: 72 miles) 42° South, 14° East
  Mersenius (crater diameter: 51 miles) 21° South, 49° West
  Metius (crater diameter: 54 miles) 40° South, 44° East
  Moretus (crater diameter: 73 miles) 70° South, 8° West
  Mutus (crater diameter: 47 miles) 63° South, 30° East
  Neper (crater diameter: 75 miles) 7° North, 83° East
  Newton (crater diameter: 85 miles) 78° South, 20° West
* Oceanus Procellarum 10° North, 47° West
  Olbers (ray crater; crater diameter: 42 miles; ray pattern diameter: 500 miles) 7°
     North, 78° West
  Orontius (crater diameter: 74 miles) 40° South, 4° West
  Palus Epidemiarum 31° South, 26° West
  Palus Nebularum 38° North, 1° East
  Palus Putredinis 27° North, 1° West
  Palus Somnii 15° North, 46° East
  Petavius (crater diameter: 110 miles) 25° South, 61° East
  Phocylides (crater diameter: 75 miles) 54° South, 58° West
  Piccolomini (crater diameter: 56 miles) 30° South, 32° East
  Pitiscus (crater diameter: 51 miles) 51° South, 31° East
* Plato (crater diameter: 63 miles) 51° North, 9° West
  Pontecoulant (crater diameter: 60 miles) 69° South, 65° East
* Posidonius (crater diameter: 63 miles) 32° North, 30° East
  Proclus (ray crater; crater diameter: 19 miles; ray pattern diameter: 400 miles)
     16° North, 47° East
* Ptolemaeus (crater diameter: 93 miles) 14° South, 3° West

Purbach (crater diameter: 77 miles) 25° South, 2° West
Pythagoras (crater diameter: 80 miles) 65° North, 65° West
Riccioli (crater diameter: 99 miles) 3° South, 75° West
Rosenberger (crater diameter: 61 miles) 55° South, 43° East
Scheiner (crater diameter: 71 miles) 60° South, 28° West
Schickard (crater diameter: 134 miles) 44° South, 54° West
Schiller (crater diameter: 112 miles) 52° South, 39° West
Schomberger (crater diameter: 52 miles) 76° South, 30° East
* Sinus Iridum 45° North, 32° West
Sinus Aestuum 12° North, 9° West
Sinus Medii 0°, 0°
Sinus Roris 54° South, 46° West
Snellius (crater diameter: 50 miles) 29° South, 56° East
Stevinus (crater diameter: 46 miles) 33° South, 54° East
Stoflerus (crater diameter: 84 miles) 41° South, 6° East
Strabo (ray crater; crater diameter: 34 miles; ray pattern diamter: 400 miles) 62° North, 55° East
Theophilus (ray crater; crater diameter: 64 miles; ray pattern diameter: 675 miles) 12° South, 26° East
* Tycho (ray crater; crater diameter: 54 miles; ray pattern diameter: 1900 miles) 43° South, 11° West
Vendelinus (crater diameter: 94 miles) 16° South, 62° East
Walter (crater diameter: 82 miles) 33° South, 1° East
Wargentin (crater diameter: 53 miles) 50° South, 60° West
Wilhelm I (crater diameter: 64 miles) 43° South, 20° West
Wurzelbauer (crater diameter: 54 miles) 34° South, 16° West

EARLY MOON WATCHERS

The Moon is not only the most visible symbol in the sky, it has long been the object of study by humans, as a measure of the passage of time as well as a celestial mystery. Early cultures, however, lacking an understanding of the heliocentric (Sun-centered) nature of the solar system, generally confined their studies to the support of religious beliefs.

Observations and records dating back thousands of years, however, show that early observers had developed an understanding of lunar cycles—many of the earliest calendars were based on the Moon, and prehistoric monuments such as Stonehenge in England and Chaco Canyon in the southwestern United States, show evidence of markings related to key solar and lunar events, including the 19-year lunar cycle (the Metonic cycle, as it is known in modern times).

Aristarchus of Samos, a Greek astronomer born circa 310 BCE, used geometry, trigonometry—and a lunar eclipse—to figure out the relative distances

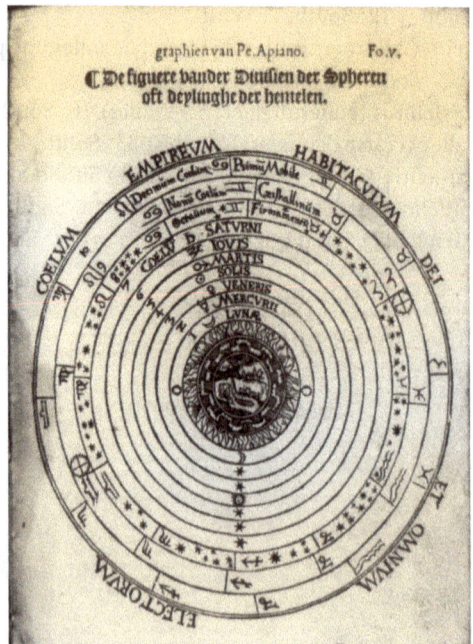

In the Middle Ages, most European cultures believed in an Earth-centered concept of the universe, with the Sun, Moon, and planets revolving around our planet. This illustration, from a book published in 1545, by by Peter Apian, shows the key revolutions as concentric rings.

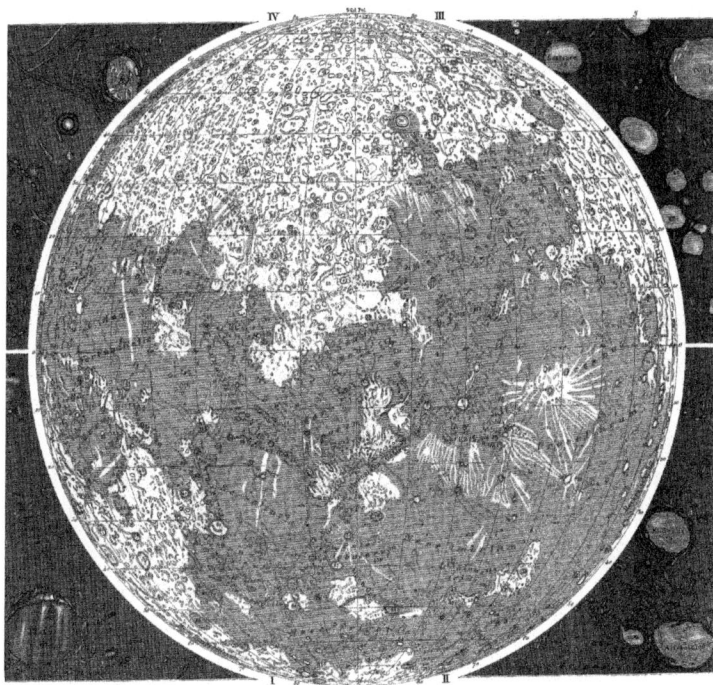

Following the earliest use of telescopes, observers have produced maps of the Moon's surface. This line engraving is from *The Iconographic Encyclopedia of Science, Literature and Art*, published in 1851 (R. Garrigue, New York), and reproduces a map first created in 1834 by Wilhelm Beer and J. H. von Mädler, German astronomers credited with the first systematic use of names for surface features.

of both the Sun and the Moon, as well as their relative sizes of one. Aristarchus influenced other Greek philosophers and his work also played a role in the development of astronomical models of the universe in India, China, and Arabic cultures.

A later development in scientific reasoning about the orbit of the Moon came from Nicolas Copernicus (1473–1543), who created a theory about the revolution of the Moon around the Earth and the planets around the Sun. Tycho Brahe, a Danish astronomer (1546–1601), believed in a geocentric (Earth-

93

centered) theory of the universe, but was responsible for developing accurate measurements of the lunar orbit, including the slight variations created from the effect of the Sun's gravity.

Johannes Kepler (1571–1630) believed, like Copernicus, that the Sun was the center of the solar system, and made pertinent observations on the Moon's orbit, the lunar effect on tides, and the nature of the lunar surface. Kepler and Galileo Galilei (1564–1642) both contributed to early lunar research by creating and refining observations of the Moon.

Along with the widespread adoption of the telescope by astronomers and scientists in the seventeenth century came the first detailed lunar maps. The first known drawing of the Moon made from observation through a telescope was from Thomas Harriot, an English astronomer and mathematician, in July 1609. Galileo made his first drawing of the Moon a few months later.

Johannes Hevelius (1611–1687) was an astronomer who specialized in the study of the lunar surface, ultimately publishing a description of the Moon that included the first standardized names for lunar features. The Hevelius nomenclature has not survived the test of time, however, being replaced by that of another astronomer from that era, Giovanni Riccioli (1598–1671). Riccioli's lists of lunar features included names of prominent landmarks, craters, and seas, many of which are still in use.

The first map of the Moon based on a system of coordinates was created by Johann Mayer (1723–1762) in 1750. The first lunar landscapes that included measurements were published in 1791 by Johann Schroeter (1745–1816). In 1835, exaggerated reports of the astronomer Sir John Herschel (1792–1871) and his discoveries about the Moon were published in a popular newspaper in New York City. These stories included supposed sightings of lunar animals, but were written without the knowledge or consent of Herschel.

Many other astronomers contributed to the growing study of the Moon, taking advantage of developments in technology and science. The first photograph of the Moon was produced on March 23, 1840, by John Draper (1811–1882). By the end of the nineteenth century, several major observatories published books of photographs of the Moon, including Lick Observatory and the Paris Observatory.

The invention and perfection of rocket-powered flight in the early part of the 20th century was the first step towards a close-up examination of the Moon. Although the scientific theory about space flight to the Moon was already well developed by the 1950s, it took a U.S.-U.S.S.R. "space race" to create the final impetus for the first unpiloted and piloted expeditions to get there.

Hundreds of pounds of moon rocks, close-up study, and huge quantities of remote sensing data have resulted from the exploration of the Moon. This information has been useful in developing new theories about the solar system and creation of the planets, and has also proven the potential of the Moon for providing huge quantities of valuable materials useful in the further exploration of space.

With a new series of lunar and planetary probes in the 1990s, interest has increased in the exploration of the Moon. One preliminary finding from the spacecraft *Clementine*, flown in 1994, indicated the possible presence of ice in deep pockets permanently shielded from sunlight and heat. In 1998, a more thorough probe conducted by the Lunar Prospector confirmed this finding, setting off a wave of speculation about the potential usefulness of this commodity for future manned Moon bases.

Along with the United States, China, Japan, the European Union, and Russia are working on or considering expanded lunar exploration, with commercial and scientific objectives in mind.

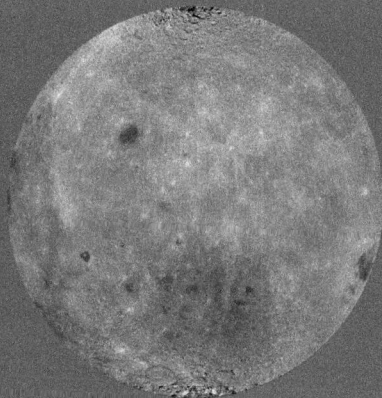

In 1994, the spacecraft *Clementine* mapped the entire surface of the Moon using special cameras. This picture of the far side is a composite made from thousands of small images, generated to record the albedo of features on the surface.

LAGRANGE POINTS

The gravitational physics of orbiting bodies produces a unique condition. Where there are examples of two large celestial bodies orbiting in relation to each other, as in the case of the Earth orbiting around the Sun, five specific points in the orbital patterns have the effect of cancelling the gravitational and centrifugal pull of the two bodies. These points are called Lagrange points, after their discovery by Joseph Louis Lagrange, a French mathematician, in 1772.

Lagrange points are potentially important spots for the future, because spacecraft, space stations, or permanent space colonies could remain in stable orbits at these locations without the need for constant fuel expenditure to maintain position.

The Lagrange points in the Earth-Moon system are also affected by additional forces from the Sun. In order to remain unaffected by these forces, objects would have to be placed into elliptical orbits around central points defined by L-4 or L-5.

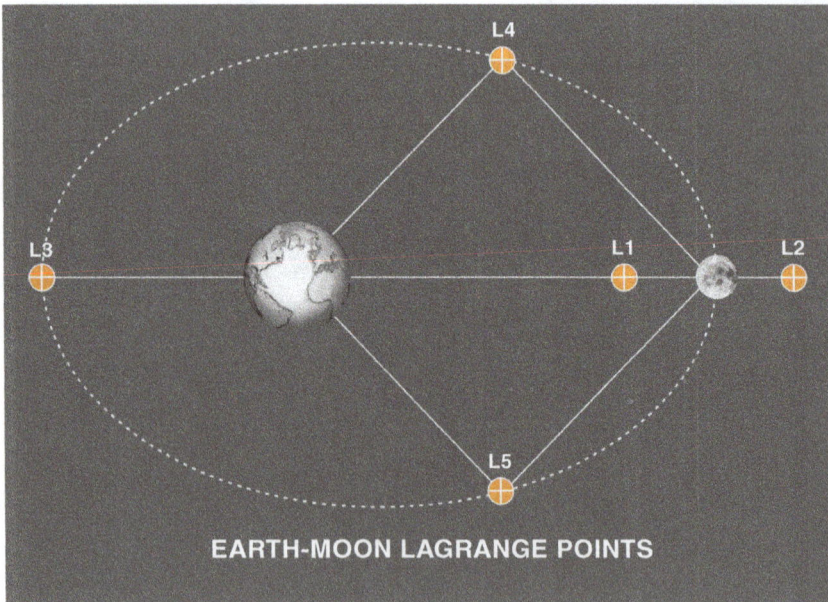

EARTH-MOON LAGRANGE POINTS

UNPILOTED MOON EXPLORATION

Pioneer 1	USA	Oct. 11, 1958	flyby at 71,300 miles
Pioneer 3	USA	Dec. 6, 1958	flyby at 66,654 miles
Luna 1	USSR	Sep. 12, 1959	landed on lunar surface
Ranger 6	USA	Feb. 1964	cameras malfunctioned
Ranger 7	USA	July 28, 1964	first pre-impact photos
Ranger 8	USA	Feb. 17, 1965	transmitted photograph
Ranger 9	USA	Mar. 21, 1965	transmitted photographs
Luna 5	USSR	May 1965	unsuccessful soft landing
Zond 3	USSR	July 18, 1965	orbited, transmitted photos
Luna 7	USSR	Oct., 1965	unsuccessful soft landing
Luna 8	USSR	Dec., 1965	unsuccessful soft landing
Luna 9	USSR	Jan. 31, 1966	landed, transmitted photos
Luna 10	USSR	Mar. 31, 1966	orbited, gamma-ray sensing
Surveyor 1	USA	May 30, 1966	landed, first color photos, data
Lunar Orbiter	USA	Aug. 10, 1966	orbited, transmitted photos, data
Luna 11	USSR	Aug. 24, 1966	orbited
Surveyor 2	USA	Sep., 1966	unsuccessful soft landing
Luna 12	USSR	Oct. 22, 1966	orbited, transmitted data, photos
Lunar Orbiter 2	USA	Nov. 7, 1966	orbited, transmitted photos, data
Luna 13	USSR	Dec. 21, 1966	landed, photos, soil sampler
Lunar Orbiter 3	USA	Feb. 4, 1967	orbited, photos, data
Surveyor 3	USA	Apr. 17, 1967	landed, tested lunar soil, data
Lunar Orbiter 4	USA	May 4, 1967	orbited, remote orbit change
Surveyor 4	USA	July 14, 1967	contact lost
Explorer 35	USA	July 19, 1967	orbited, magnetic fields
Lunar Orbiter 5	USA	Aug. 2, 1967	orbited, data
Surveyor 5	USA	Sep. 8, 1967	landed, soil experiments, data
Surveyor 6	USA	Nov. 7, 1967	landed, transmitted photos, data
Surveyor 7	USA	Jan. 6, 1968	landed, tested soil, photos, data
Luna 14	USSR	Apr. 1968	orbited
Zond 5	USSR	Sep. 14, 1968	orbited

Zond 6	USSR	Nov. 10, 1968	flyby, returned film to Earth
Luna 15	USSR	July 1969	landed
Zond 7	USSR	Aug., 1969	flyby, returned film to Earth
Luna 16	USSR	Sep. 12, 1970	landed, returned samples to Earth
Zond 8	USSR	Oct. 1970	flyby, returned film to Earth
Luna 17	USSR	Nov. 10, 1970	landed, remote vehicle, soil tests, TV
Luna 18	USSR	Sep. 1971	unsuccessful soft landing
Luna 19	USSR	Sep. 28, 1971	orbited, remote sampling
Luna 20	USSR	Feb. 1972	landed, returned samples to Earth
Luna 21	USSR	Jan. 8, 1973	landed, remote vehicle, returned to Earth with soil samples
Luna 22	USSR	May 29, 1974	orbited
Luna 23	USSR	Nov. 1974	damage during landing
Luna 24	USSR	Aug. 9, 1976	landed, returned samples to Earth
Muses A	JAPAN	Jan., 1990	orbited
Galileo	USA	Oct. 18, 1990	fly-by, remote sampling, photos
Clementine	USA	Jan. 25, 1994	orbited, remote measurement, mapping
AsiaSat 3/HGS-1	Hong Kong	Dec. 24, 1997	flyby
Lunar Prospector	USA	Jan. 7, 1998	orbited, remote sampling, mapping
SMART 1	USA	Sep. 27, 2003	orbited
Kaguya (SELENE)	Japan	Sep. 14, 2007	orbited
Chang'e 1	China	Oct. 24, 2007	orbited
Chandrayaan-1	India	Oct. 22, 2008	orbited
LRO	USA	June 17, 2009	orbited
LCROSS	USA	June 17, 2009	orbited, impacted
Change'e 2	China	Oct. 1, 2010	orbited
GRAIL	USA	Sep. 10, 2011	orbited
LADEE	USA	Sep. 6, 2013	orbited
Change'e 3	China	Dec. 1, 2013	landed, explored

LUNAR TRAJECTORIES

Flight time to the Moon: 36 to 72 hours

Trajectory for lunar landing

Trajectory for lunar orbit

THE APOLLO PROGRAM: MEN ON THE MOON

MISSION	LAUNCH	DURATION (days, hours, minutes)	CREW
Apollo 7	Oct. 11, 1968	10d 20h 9m	Cunningham, Eisele, Schirra

Orbital tests around Earth of Apollo command and service module

| Apollo 8 | Dec. 21, 1968 | 6d 3h 0m | Anders, Borman, Lovell |

First flight to Moon; first orbit around Moon; inclination about 13°, perilune 110 km

| Apollo 9 | Mar. 3, 1969 | 10d 1h 0m | McDivitt, Schweickart, Scott |

Orbital tests around Earth, first flight of complete Apollo spacecraft

| Apollo 10 | May 18, 1969 | 8d 0h 3m | Cernan, Stafford, Young |

Orbital tests around Moon of complete Apollo spacecraft; partial descent to surface by lunar module; inclination about 1°, perilune 110–15 km)

| Apollo 11 | July 16, 1969 | 8d 3h 18m | Aldrin, Armstrong, Collins |

First Moon landing; first walk on Moon (Armstrong); Mare Tranquillitatis, Statio Tranquillitatis, lat 0.7° N, long 23.4° E

| Apollo 12 | Nov. 12, 1969 | 10d 4h 36m | Bean, Conrad, Gordon |

Landing; surface exploration; Oceanus Procellarum, lat 3.2° S, long 23.4° W

| Apollo 13 | Apr. 11, 1970 | 5d 0h 1m | Haise, Lovell, Swigert |

Flyby; mission aborted in 3rd day

| Apollo 14 | Jan. 31, 1971 | 9d 0h 1m | Mitchell, Roosa, Shepard |

Landing; surface exploration; Fra Mauro highlands, lat 3.7° S, long 17.5° W

| Apollo 15 | July 26, 1971 | 12d 7h 11m | Irwin, Scott, Worden |

First use of Lunar Rover; first continuous color TV broadcast of moon walk; extensive scientific study of lunar surface; Palus Putredinis, Apenninus-Hadley region, lat 26.1° N, long 3.7° E

| Apollo 16 | Apr. 16, 1972 | 11d 1h 51m | Duke, Mattingly, Young |

Landing; surface exploration; Descartes highlands, lat 9.0° S, long 15.5° E

| Apollo 17 | Dec. 7, 1972 | 12d 13h 1m | Cernan, Evans, Schmitt |

Landing; first geological study of lunar surface (Schmitt); Taurus-Littrow valley, lat 20.2° N, long 30.8° E

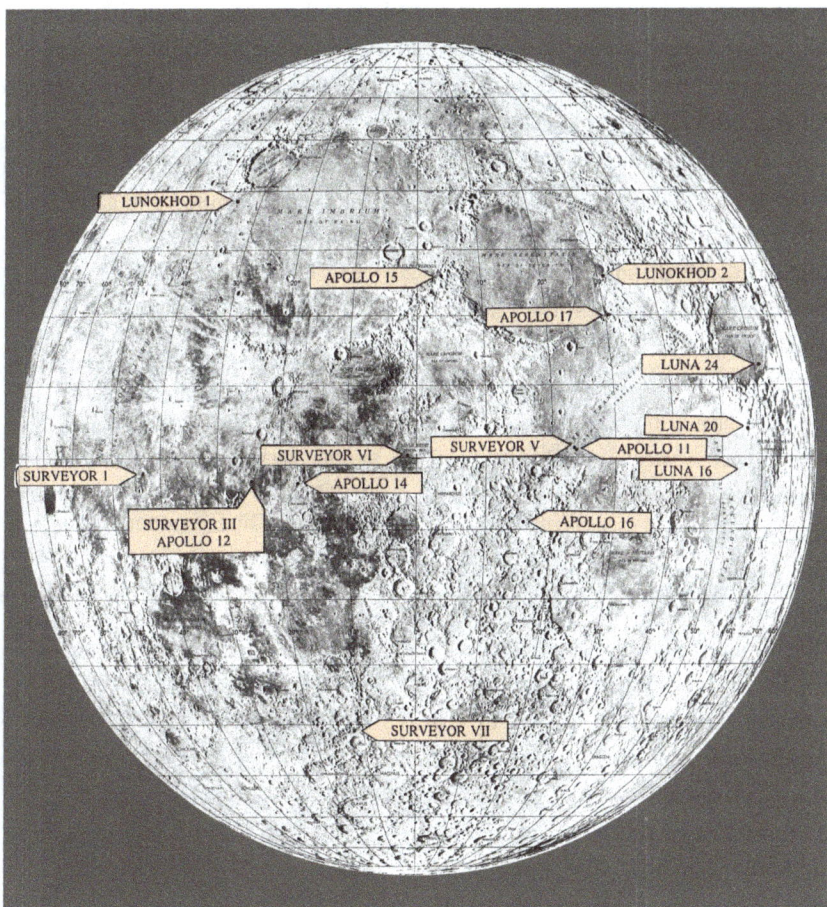

LUNAR LANDING SITES

Graphic modified from an original map courtesy of the Defense Mapping
Agency/Lunar and Planetary Institute.

THE CREATION OF THE MOON

Despite an increasing depth of knowledge about the composition of the Moon, scientists have as yet no provable theory about its origin. One general agreement, however, is that the Moon was formed at least 4.5 billion years ago during the creation of the solar system. One theory suggests that it was formed near to, but separate from the Earth; another suggests that it was formed some distance away but was pulled into orbit around the Earth by gravitational attraction over time.

One of the most accepted theories is that a large object such as an asteroid impacted the Earth, throwing off enough vapor and material to form the Moon. During the first phase of the Moon's existence, forces caused by the cooling molten material produced many of its existing features.

The largest visible features on the surface were created by violent collisions with asteroids that may have been as large as the state of Delaware. The interior of the Moon is believed to have originally been more solid, with melting caused by radioactive decay. The last violent surface activity on the Moon probably happened more than 3 billion years ago, with volcanic-like eruptions and flooding of lava across the surface.

MOONQUAKES

Much is still unknown about the interior of the Moon, but scientists believe that there is a core of molten or partly molten material. There are different layers in the structure of the moon; shifting of the layers produces tremors similar to earthquakes. These moonquakes are usually very weak—many of them release no more energy than a firecracker. Other moonquakes are caused by the impact of meteorites on the surface and some occur at regular intervals during a lunar cycle, suggesting that gravitational forces from the Earth, similar to ocean tides, may cause movement within the body of the Moon.

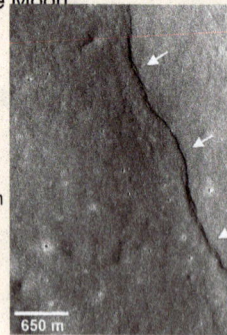

RIGHT A thrust fault at the side of an impact crater on the far side of the Moon, photographed from orbit.

Photo courtesy of NASA/Goddard/ Arizona State U./ Smithsonian.

650 m

MOON ROCKS

Examination of the rocks brought back by the Apollo astronauts is still underway, decades after the missions were concluded. Scientists have discovered many interesting features about the composition and origin of the Moon from these rocks, most of them formed from cooling lava and therefore igneous in nature. Some of the rocks are similar to basalt, which is found on Earth. Samples of lunar basalt were collected in low areas of the surface that are observed as maria from the Earth.

Rocks from higher regions of the Moon are also igneous, and are referred to as gabbro, norite, and anorthosite, similar to rocks of the same names on Earth. Although moon rocks have some characteristics similar to Earth rocks, they are recognizably different because of the complete lack of water; when present, water has a noticeable effect on minerals in rocks. Moon rocks also exhibit crystals of metallic iron that occur because of the lack of free oxygen. Lunar minerals include feldspar, olivine, pyroxine, ilmenite, plagioclase, and troilite.

Material on the surface of the Moon is referred to as regolith or lunar soil, but it has no organic content. Lunar soil forms a layer from 3 to 60 feet (1 to 20 meters) deep on the surface. This layer is composed of rocks and powder but was not formed by eroding forces such as wind or water. Instead, lunar soil was created over a period of billions of years by the continuous bombardment of meteorites. Larger meteorites form visible craters that can be seen from the Earth; smaller, virtually invisible craters are formed by particles of cosmic dust. The smallest craters are only 1/25,000 inch (1/1000 millimeter) in diameter.

RIGHT One batch of rock samples returned from the Apollo 11 mission. Photo courtesy of NASA/JSC.

MOON CALENDARS

Most ancient civilizations based their calendars on the lunar cycle. The highly visible phase changes of the Moon made observations and accurate date projections relatively easy compared to the Sun, which repeats itself over a longer period, a year. There is, however, a built-in problem with this method. Most of the seasonal variations in climate are linked to the solar year, and there is no even number of lunar months which equals one solar year. Therefore calendars based on lunar months are always out of step with the seasons, and extra days or months must be added periodically to make them practical. This procedure is referred to as intercalation.

The ancient Babylonian calendar was based on a lunar month that began when the first crescent Moon was visible. There were 12 months in every year, with months alternating between 30 and 29 days long. This lunar calendar would have been out of phase with the solar year, which is almost 12 days longer, except that the Babylonians added special intermediate (or intercalary) months seven times in every calendar cycle. The extra days in these special months made the cycle come out even (start repeating itself) every 19 years.

The 19-year cycle was also adopted by other cultures, including the Greeks, and is also used in the present-day Jewish calendar. Each of these cycles is the equivalent of 235 lunar months, or lunations.

The Egyptian civilization was heavily influenced by the annual flooding of the Nile River. The time of flooding every year coincided with the rising of the star Sirius close to the Sun, so the Egyptians created a lunar calendar that had its first month begin when the new moon occurred after the rising of

CHINESE LUNAR MONTHS

1	Holiday Moon
2	Budding Moon
3	Sleepy Moon
4	Peony Moon
5	Dragon Moon
6	Lotus Moon
7	Moon of Hungry Ghosts
8	Harvest Moon
9	Chrysanthemum Moon
10	Kindly Moon
11	White Moon
12	Bitter Moon

Sirius. This lunar calendar featured 12 months, each one 29½ days long. Extra months were included occasionally to keep pace with the solar year. The Egyptians were very advanced in observations of celestial cycles and eventually created a solar calendar to replace the lunar one.

Early Greeks developed and relied on lunar calendars that were organized and maintained by each separate city or town. Beginning about the sixth century BCE, Greek astronomers and mathematicians created more organized lunar calendars based on 19-year cycles, but they eventually switched to the Roman calendar, which was based on the Sun.

The Romans originally devised a lunar calendar but reorganized it during the reign of Julius Caesar. Caesar added extra days to the lunar calendar to keep it from getting out of step with the solar year; this new calendar was first used in 45 BCE. This calendar was referred to as the Julian calendar and was widely used in western countries until a further reform was organized in 1582. At that time, the rapid spread of Christianity necessitated the fixation of religious holidays based on the celebration of Easter. Pope Gregory XIII was responsible for

RELIGIOUS MONTHS

MONTH	JEWISH	MOSLEM
1	Tishri	Muharram
2	Heshvan	Safar
3	Kislev	Rabi'u'l-Avval
4	Tevet	Rabi'u'l-Thani
5	Shevat	Jumadiyu'l-Avval
6	Adar	Jumadiyu'th-Thani
7	Nisan	Rajab
8	Iyyar	Sha'ban
9	Sivan	Ramadan
10	Tammuz	Shavval
11	Av	Dhi'l-Qa'dih
12	Elul	Dhi'l-Hijjih

this reform, and the new calendar, still in use today, is called the Gregorian calendar. The Gregorian calendar measures a year at 365 days (an average of 365.2425 days if leap year is included) and comes very close to matching the solar cycle. Every 400 years, the total difference between the two is only a few hours. By comparison, the error with the Julian calendar was about three days for the same period.

The Gregorian calendar that we use today is based on a requirement within the Christianity to determine the correct date for Easter every year. Easter is considered the beginning of the Christian calendar and all Christian holidays and special days are figured from that date. In modern times, Easter is determined by a fixed set of rules that modify the actual lunar cycle in order to keep

HAWAIIAN LUNAR DAYS

AGE OF THE MOON	HAWAIIAN NAME	AGE OF THE MOON	HAWAIIAN NAME
1	Hilo	15	Hoku
First appearance crescent moon		16	Mahealani
2	Hoaka	17	Kulu 18 Laau-ku-kahi
3	Ku-Kahi	18	Laau-ku-lua
4	Hu-lua	20	Laau-pau
5	Hu-kolu	21	Ole-ku-kahi
6	Hu-pau	*Last quarter moon*	
7	Ole-ku-kahi	22	Ole-ku-lua
First quarter moon		23	Ole-pau
8	Ole-ku-lua	24	Kaloa-ku-kahi
9	Ole-ku-kolu	25	Kaloa-ku-lua
10	Ole-ku-pau	26	Kaloa-pau
11	Huna	27	Kane
12	Mohalu	28	Lono
13	Hua	29	Mauli ("moon is fainting")
14	Akua	30	Muku ("moon cut off by the Sun")
Full moon			

the date between March 22 and April 25. If the real, observable lunar cycle were used, the date for Easter could vary much more.

In practice, a fixed date for the spring equinox is set at March 21 and Easter is determined by finding the first Sunday following the full moon that occurs on or just after March 21. This full moon date is not determined from the observable cycle but by special religious tables that vary slightly from the real times. Not by coincidence, these tables are based on the same 19-year lunar cycle which was used by the Babylonian, Greek, and Jewish calendars.

Among those calendars that rely on the Moon, lunar months may run from full moon to full moon or new moon to new moon, depending on tradition. In southern India, the month begins with the new moon; in other parts of the country, the full moon signals the beginning.

The Jewish calendar was first created around the sighting of the crescent moon, which determined the beginning of each of 12 lunar months. At occasional intervals over the years, the twelfth month was repeated to synchronize

HAWAIIAN LUNAR MONTHS

1 Moakali'i (also Hui-hui, refers to the Pleiades)

2 Ka'elo ("Drenching Month," refers to Betelgeuse)

3 Kaulua ("Pit of Sacrifice," named for Sirius)

4 Nana

5 Welo

6 Ikiiki ("Warm and sticky")

7 Ka'aona

8 Hinaia-'ele'ele

9 Hili-na-ehu ("Sea-borne mists")

10 Wehewehe

11 Hili-na-ma

12 Ikuwa (also Kauka-Malama, Welehu, Welehu-lua, or Welehu II, an extra month used to compensate for the rising of the Pleiades when it comes late in the year)

the calendar with the seasons. Until the fourth century CE, there were various versions of this calendar in use. Jews in some countries began their year with the month of Nisan in the spring; other countries used the month of Tishri in the fall. The modern Jewish calendar was adopted in the fourth century CE and relies on fixed sequences of months based on a 19-year cycle of lunations. This 19-year cycle is the same one that was used by the Babylonians.

The new calendar determines the beginning of the year from the time of the new moon in the month of Tishri (occurring in the fall) but uses special rules to keep important religious days from falling on the wrong days. The Jewish year can have 353, 354, or 355 days, with occasional "leap years" of 383, 384, or 385 days. For every 19-year cycle, there are 12 regular years and seven "leap years."

The Muslim calendar is also based on lunar months but uses a cycle of 33 years. There are 12 lunar months in the Moslem calendar, with lengths alternating between 29 and 30 days. The calendar is fixed instead of relying on actual observation of the Moon, but it is only off from the actual lunar cycle by one day every 2,500 years.

INDIAN ASTROLOGY

Traditional Indian astrology is based on 27 lunar months, called nakshatra ("mansions"). Each of these months begins with the rising of the Moon at a particular point on the ecliptic.

1	Aswini		15	Swati
2	Bharani		16	Visakha
3	Krittika		17	Anuradha
4	Rohini		18	Jyeshtha
5	Mriga		19	Mula
6	Ardra		20	Purva-shadha
7	Punarvasu		21	Uttara-shadha
8	Pushya		22	Stravana
9	Ashlesha		23	Dhanishtha
10	Magha		24	Satataraka
11	Purva-phalguni		25	Purva-bhadrapada
12	Uttara-phalguni		26	Uttara-bhadrapada
13	Hasta		27	Revati
14	Chitra			

In practice, however, many Muslims base their religious holidays on actual observation of the first crescent moon but rely on a modified lunar calendar—or western Gregorian calendar—for civil activities. The important dates are figured from the first observation of the crescent moon, and these religious days begin at sunset. The first year of use for the Moslem calendar was 622 CE, the year Mohammed left Mecca.

Most of India also uses the Gregorian calendar for day-to-day activities, but Hindu religious dates are based on a very old lunar calendar. The traditional Hindu calendar dates back to at least 1000 BCE. It is organized around 12 lunar months of 27 or 28 days each. About every 60 months, an extra month is added to keep it in step with the solar year. Each month is also divided into two parts, corresponding to the waxing and waning phases. In addition, the Hindu calendar also uses a complicated system relating Earth's solar orbit with that of Jupiter's, and uses a unique division of daily time into *vipalas* (0.4 seconds each) and *ghatikas* (24 minutes each).

The oldest lunar calendar currently in use is Chinese. At least one estimate puts the first year of use of the traditional Chinese calendar at 2698 BCE. This calendar is used to determine dates of traditional festivals and special religious days, but it is no longer used for daily or civil activities. It was banned in 1912 when the Republic of China was formed and the traditional name for this calendar system was changed to Nung Li, meaning Farmers' Calendar. The Chinese government and most of the country now rely on the western Gregorian calendar for day-to-day activities.

The traditional Chinese calendar is based on 12 lunar months, with each month having either 29 or 30 days. Occasionally a special intermediate month was added to keep the lunar months in line with the solar year. The calendar runs on a 60-year cycle, at which time it begins repeating itself. Inside this big cycle, however, are five smaller cycles, each lasting 12 years. These 12 years are named after animals. The Chinese new year starts with the first new moon that occurs after the Sun has entered the constellation of Aquarius. In practice, New Year's Day in the Chinese calendar can occur from January 20 to February 19.

The traditional culture of the Hawaiian Islands also has a lunar calendar, one based on the rising of stars. Each lunar month is called a *mahina* and begins when the Sun sets on the first day after the new moon. Months are either 29 or 30 days in length and each day in a month is named for the appearance of the Moon on that day or a period of time after a particular phase. The first lunar month begins with the rising of the Pleiades constellation in the east ("Hakali'i" or "Hui-hui" in Hawaiian).

In modern times, some major religions still base calendars on a lunar timetable. Both the Hebrew and Islamic calendars, for example, use the Moon as a basis for calculating the passage of time. In the case of Muslim tradition, orthodox or fundamentalist members believe that the beginning of each lunar month cannot begin until the first crescent moon has actually been seen by human eyes. But although the practice of looking for the first crescent moon continues, modern crescent sighters sometimes rely on computer programs, cell phones, and the Internet to increase the effectiveness of their efforts (see "First Sighting," page –).

THE FIRST WORD

The modern English word "moon" gradually developed from earlier versions. In Old English, the word was "mona" and in Middle English, "mone" or "moone." The root of these comes from Latin, the word "mensis," and before that, in Greek, it was "men." Parelleling the development in English, in old Nordic the word was "mani" and in old high German and Old Saxon, it was "mano," (pronounced mayno). One of the earliest published references in Old English dates to 698 AD. The word "month" has a parallel past, beginning with the same roots as "moon."

AMERICAN SIGN LANGUAGE

The sign for "Moon" in ASL is formed by making the shape of the letter "C" with the right hand and holding it up to the right eye.

FOREIGN MOONS

AFRIKAANS maan
ARABIC kamar or qamar
BASQUE hilargi
BRETON loar
CARIB nu'nû
CATALAN lluna
CEBUANO bulan, buwan
CHECHEN butt
CHINESE yueh
CHINESE (PINYIN) yuèqiú
CHINESE MANDARIN yuèliang
DANISH måne
DUTCH maan
EGYPTIAN pooh
ESKIMO tatkret
ESPERANTO luno
ETHIOPIAN sin (or ilmuqah)
FIJIAN vula
FINNISH kuu
FRENCH lune
GAELIC gealach
GERMAN mond
GREEK selina, mena
GUJARATI chandra
HAITIAN CREOLE lalin
HAUSA wata
HEBREW yaréakn
HINDI-URDU chad
HUNGARIAN hold
HINDUSTANI chandra
ICELANDIC tungli
INDONESIAN bulan
IRISH GAELIC gealach, luan
JAPANESE hyourin, tsuki
JIVARO nantu
KOREAN tal
KURDISH meh, hïv

LAO pa: cha:n
LATIN luna
LATVIAN meness
LITHUANIAN menu
MACEDONIAN mu:n
MALAY bulan
MAYAN luunaa
MONGOLIAN sar
NEPALI candrama, jun
NIUEAN mahina
NORWEGIAN måne
PALAUAN búil
PANJABI cann
PERSIAN mah
PERUVIAN INDIAN sin
PHILIPINO buwán
POLISH ksiezic
PORTUGESE lua
PULAAR lewru
PYGMY pe
ROMANIAN luna
RUSSIAN luna
SAMOAN masina
SANSKRIT mas, chandraḥ
SERBO-CROATIAN mjesec
SLOVAK mesiac, luna
SOMALI dayax
SPANISH luna
SWAHILI mwezi
SWEDISH måne
THAI prá-jun
TIBETAN dah-wah
TIV uwer
TURKISH ay, mehtap
WELSH lleuad, looer
YIDDISH levone
YORUBA òṣupá

NATIVE AMERICAN MOONS

ATAKAPA iti', yil
BILOXI ina
BLACKFOOT natósi, ki'sómm
CHOCTAW hạshi
COCHITI tâ'waṭạ
CHEROKEE nvda
CREE tipiska'wi-pi'sim
DAKOTA SIOUX wi
DELAWARE nipáhum
HAIDA qoñ
HAWAIIAN mahina
HOPI muuyaw
HUPA xatLe wha
INUKTITUT ESKIMO tatkret
JEMEZ p̂â
KOASATI nithahasí
KWAKIUTL ^{E}mEku'la
MENOMINI típä'kē'so
MICMAC kisuhs, depkunoosat
MISKITO kati
MUSKOGEE hvréssē
NARRAGANSETT nanepaùshat
NATICK nepauzshad
NAVAJO ooljéé, tł'ééhona'ái
NEZ PERCE hí-semtuks
OFO i'la
OJIBWAY ne-bah-geeses, deebee-kee-zeis
ONONDAGA garáchqua
OSAGE wa-ḳo^n'-da ho^n-do^n
PAPAGO-PIMA marshad
POTAWATOMI tpukisIs
SHOSHONE
TAOS p̂aenâ
TLINGIT dîs
ZUNI jáunanne

SIGN LANGUAGE

Many of the Indian cultures in the western regions of North America shared a common language of hand signs. Using this sign language, one way to signify the word "moon" was as follows. First, the sign was made for the word "night," by extending both hands in front of the body with the palms down and a few inches apart. The right hand is held a little higher than the left hand and both hands are turned inward, crossing over one another. Next, the right hand was used to form the letter "C" by curving the index finger and thumb, the ends about an inch apart, and the other fingers closed. This hand is then raised above the head. The names of specific months could also be signalled by adding another sign to describe the activity traditionally associated with that month.

AMERICAN FULL MOONS

American colonists brought many European traditions with them when they settled this country. Among those traditions was the naming of full moons. In Europe, these traditional names were often connected to religious—mostly Christian—dates. In the New World, the naming of full moons was also influenced by the traditions already established in northeastern North America by Native American tribes, mostly Algonquin.

Tribes in other parts of the country often had different names for the moons, usually related to natural changes caused by the seasons. In some cases, twelve distinct names were used but there were tribes who used no more than six names for an entire year, repeating terms to make up the difference. Tribes also shared moon names and sometimes changed the name of a moon from one year to the next.

Compounding this convention, even among members of a single tribe, scattered groups might sometimes use different names for the same moon or the same name for different moons. The Santee band of the Dakotah Sioux, for example, traditionally called the full moon in September, "Moon When the Horns Are Broken Off," but the same name was used for the full moon in December by the Teton band of the same tribe.

In modern western cultures, many people continue to use the traditional colonial names for at least a few key full moons, the harvest moon being the most common. However, because the calendar dates of full moons continually regress—move backward relative to the calendar from year to year—there are periods when the standard full moon names do not match up with calendar months. The full moon closest to the fall equinox, for example, can occur in early October, "bumping" the following full moons back.

In reality, the names of the full moons follow a convention, not a rigid set of rules. This has been the case for hundreds of years, as represented by almanacs published in previous centuries that sometimes disagreed on which moon name was appropriate for which month.

JANUARY The first full moon after the winter solstice or the first full moon after Yule.

Colonial American	Winter Moon (also Yule Moon)
Algonquin	Wolf Moon (also Old Moon)
Cherokee	Month of the Cold Moon
Cheyenne	Hoop and Stick Game Moon
Choctaw	Cooking Moon
Dakotah Sioux	Moon of the Terrible
Haida	Younger Moon
Ildefonso	Ice Moon
Klamath	Moon of the Little Finger's Partner
Kutenai	Naktasu Moon (no translation)
Laguna	Lizard Cut Moon
Lakota Sioux	Moon of Frost in the Teepee
Micmac	Frost-Fish Moon
Mohawk	The Big Cold
Natchez	Cold Meal Moon
Nez Perce	Cold Weather Moon
Ojibway	Great Spirit Moon
Osage	Hunger Moon
Oto	The Little Young Bear Comes Down the Tree
Ponca	Snow Thaws
San Juan	Ice Moon
Taos	Man Moon
Tlingit	Goose Moon
Wisham	Her Cold Moon
Zuni	Trees Broken Moon (same as July)

FEBRUARY The second full moon of the year, associated with the middle of winter.

Colonial American	Trapper's Moon, Snow Moon, Storm Moon
Algonquin	Snow Moon (also Hunger Moon)
Cherokee	Month of the Bony Moon
Cheyenne	Big Hoop and Stick Game Moon

Choctaw	Little Famine Moon
Dakotah Sioux	Moon of the Raccoon, Moon When Trees Pop
Haida	Elder Moon
Kutenai	Black Bear Moon
Laguna	Yamuni Moon (Yamuni is an edible root)
Lakota Sioux	Moon of the Dark-red Calves
Micmac	Snow-Blinding Moon
Mohawk	Lateness
Natchez	Chestnut Moon
Nez Perce	Budding Time
Ojibway	Sucker Moon
Osage	Light of Day Returns Moon
Oto	Raccoon's Rutting Season
Ponca	Moon When the Ducks Come Back to Hide
San Ildefonso	Wind Moon
San Juan	Coyote Frighten Moon
Taos	Winter Moon
Tewa	Moon When the Coyotes are Frightened
Tlingit	Black Bear Moon
Wisham	Shoulder Moon
Zuni	No Snow on Trails Moon (same as August)

MARCH The last full moon before the spring equinox.

Colonial American	Fish Moon (also Worm Moon, Sap Moon, Crow Moon, Lenten Moon, Chaste Moon)
Algonquin	Worm Moon (also Crow Moon, Crust Moon, Sap Moon)
Cherokee	Month of the Windy Moon
Cheyenne	Light Snow Moon (also Dusty Moon)
Choctaw	Big Famine Moon
Dakotah Sioux	Moon When Eyes Are Sore from Bright Snow
Delaware	Moon when the Juice Drips From the Trees
Haida	Tahet Moon (Tahet is a type of salmon)
Kutenai	Earth Cracks Moon
Laguna	Schamu Moon (Schamu is a local plant)

Lakota Sioux	Moon of Snow-blindness
Micmac	Spring Moon
Mohawk	Much Lateness
Natchez	Deer Moon
Nez Perce	Flower Time
Ojibway	Breaking Up of Snow Shoes Moon
Osage	Just Doing That Moon
Oto	Big Clouds Moon
Ponca	Sore-Eyes Moon
San Ildefonso	All Leaf Split Moon
San Juan	Lizard Moon
Taos	Wind Strong Moon
Tlingit	Moon When Sea Flowers and All Other Things Under the Sea Begin to Flower
Wisham	The Seventh Moon (also Long Days Moon)
Zuni	Little Sandstorm Moon (same as September)

APRIL The first full moon after the spring equinox.

Colonial American	Planter's Moon (also Easter Moon, Pink Moon, Grass Moon, Egg Moon, Seed Moon)
Algonquin	Pink Moon (also Sprouting Grass Moon, Egg Moon, Fish Moon)
Cherokee	Month of the Flower Moon
Cheyenne	Spring Moon
Choctaw	Wildcat Moon
Dakotah Sioux	Moon When Geese Return in Scattered Formations, Moon to Go Paddling
Haida	Ketkakaitash Moon (no translation)
Illinois	Do Nothing Moon
Kutenai	Deep Water Moon
Laguna	Sticky Mud Plant Moon
Lakota Sioux	Moon of Grass Appearing
Micmac	Egg-Laying Moon
Mohawk	Budding Time
Natchez	Strawberries Moon
Nez Perce	Kaket Time (Kaket is an edible root)

Ojibway	Boiling Down of Sap Moon
Osage	Planting Moon
Oto	Little Frogs Croak Moon
Ponca	Rains Moon
San Ildefonso	Leaf Spread Moon
San Juan	Leaf Split Moon
Taos	Ashes Moon
Tlingit	Real Flower Moon
Wisham	The Eighth Moon
Zuni	Great Sandstorm Moon (same as October)

MAY The fifth full moon of the year.

Colonial American	Milk Moon (also Mother's Moon, Hare Moon)
Algonquin	Flower Moon (also Corn Planting Moon, Milk Moon)
Cherokee	Month of the Planting Moon
Cheyenne	Time When the Horses Get Fat
Choctaw	Panther Moon
Dakotah Sioux	Moon to Plant, Moon When Leaves Are Green
Haida	Salmonberry Bird Moon
Kutenai	Deep Water Moon
Laguna	Loam Plant Moon
Lakota Sioux	Moon of the Shedding Ponies
Micmac	Young Seals Moon
Mohawk	Time of Big Leaf
Natchez	Little Corn Moon
Nez Perce	Kouse Bread Time
Nunamiut Eskimos	Moon When the Ice Goes Out of the Rivers
Ojibway	Budding Moon
Osage	Little Flower Killer Moon
Oto	To Get Ready for Plowing and Planting
Ponca	Summer Begins Moon
San Ildefonso	Planting Moon
San Juan	Leaf Tender Moon
Taos	Corn Planting Moon

Tlingit	Moon When People Know That Everything Is Going to Grow
Wisham	The Ninth Moon
Zuni	Moon No Name Moon (same as November)

JUNE The last full moon before the summer solstice.

Colonial American	Rose Moon or Strawberry Moon (also Stockman's Moon, Honey Moon, Hot Moon, Flower Moon, Dyad Moon)
Algonquin	Strawberry Moon
Cherokee	Month of the Green Corn Moon
Cheyenne	Moon When the Buffalo Bulls are Rutting
Choctaw	Windy Moon
Dakotah Sioux	Moon When June Berries Are Ripe
Haida	Berry Ripening Season Moon
Kutenai	Ripening Strawberries Moon
Laguna	Corn Moon
Lakota Sioux	Moon of Making Fat
Micmac	Leaf-Opening Moon
Mohawk	Ripening Time
Natchez	Watermelons Moon
Nez Perce	Salmon Fishing Time
Ojibway	Strawberry Moon
Osage	Buffalo Pawing Earth Moon
Oto	Hoeing Corn Moon
Ponca	Hot Weather Moon
San Ildefonso	Flower Moon
San Juan	Leaf Dark Moon
Taos	Corn Tassle Appear Moon
Tlingit	Moon of the Salmon
Wisham	Rotten Moon
Zuni	Turning Moon (same as December)

JULY   The full moon after the summer solstice.

Colonial American	Summer Moon (also Buck Moon, Thunder Moon, Hay Moon, Mead Moon)
Algonquin	Buck Moon (also Thunder Moon)
Cherokee	Month of the Ripe Corn Moon
Choctaw	Crane Moon
Dakotah Sioux	Moon When Chokeberries Are Red, Middle of the Summer Moon
Haida	Killer Whale Moon
Kutenai	Ripening Service Berries Moon
Laguna	Corn Tassle Moon
Lakota Sioux	Moon when the Cherries are Ripe
Micmac	Sea-Fowl Shed Feathers
Mohawk	Time of Much Ripening
Natchez	Peaches Moon
Nez Perce	Red Salmon Time
Ojibway	Raspberry Moon
Osage	Buffalo Breeding Moon
Oto	Buffalo Rutting Season Moon
Pima	Moon of the Giant Cactus
Ponca	Middle of Summer Moon
San Ildefonso	Rain Moon
San Juan	Ripe Moon
Taos	Sun House Moon
Tlingit	Moon When Everything Is Born
Wisham	Advance in a Body Moon
Zuni	Trees Broken Moon (same as January)

AUGUST   In most years, the last full moon of the year.

Colonial American	Dog Day's Moon (also Woodcutter's Moon, Sturgeon Moon, Green Corn Moon, Grain Moon, Wort Moon)
Algonquin	Sturgeon Moon (also Red Moon, Green Corn Moon)
Cherokee	Moon of the End of the Fruit Moon

Cheyenne	Time When the Cherries are Ripe
Choctaw	Women's Moon
Dakotah Sioux	Moon When All Things Ripen
Haida	Collect Food for Winter Moon
Kutenai	Berries Ripen Even in the Night Moon
Laguna	Yamoni Moon (Yamoni is an immature ear of corn)
Lakota Sioux	Moon When the Cherries Turn Black
Micmac	Young Birds Are Full-Fledged
Mohawk	Time of Freshness
Natchez	Mulberries Moon
Nez Perce	Summer Time
Ojibway	Blueberry Moon
Osage	Yellow Flower Moon
Oto	All the Elk Call Moon
Ponca	Corn Is in Silk Moon
San Ildefonso	Wheat Cut Moon
San Juan	Wheat Cut Moon
Taos	Autumn Moon
Tlingit	Moon When All Kinds of Animals Prepare Their Dens
Wisham	Blackberry Patches Moon
Zuni	No Snow on Trails Moon (same as February)

SEPTEMBER The full moon closest to the fall equinox. It can also fall as late as the first week in October.

Colonial American	Harvest Moon (also Fruit Moon, Dying Grass Moon, Barley Moon)
Algonquin	Harvest Moon
Cherokee	Month of the Nut Moon
Cheyenne	Cool Moon
Choctaw	Mulberry Moon
Dakotah Sioux	Moon When Wild Rice Is Stored for Winter Use
Haida	Salmon Spawning Moon
Kutenai	Ripe Choke Cherries Moon

Laguna	Corn in the Milk Moon
Lakota Sioux	Moon when the Calves Grow Hair (also Moon of the Black Calves and Moon when the Plums are Scarlet)
Micmac	Moose-Calling Moon
Mohawk	Time of Much Freshness
Natchez	The Great Corn Moon
Nez Perce	Spawning Salmon Time
Ojibway	Wild Rice Moon
Osage	Deer Hiding Moon
Oto	Spider Web on the Ground at Dawn Moon
Paiute	Moon Without a Name
Ponca	Moon When the Elk Bellow
San Ildefonso	All Ripe Moon
San Juan	All Ripe Moon
Taos	Leaf Yellow Moon
Tlingit	Small Moon
Wisham	Her Acorns Moon
Zuni	Little Sandstorm Moon (same as March)

OCTOBER The first full moon after the Harvest Moon.

Colonial American	Hunter's Moon (also Blood Moon)
Algonquin	Hunter's Moon
Cherokee	Month of the Harvest Moon
Cheyenne	Moon When Water Begins to Freeze on the Edge of the Stream
Choctaw	Blackberry Moon
Dakotah Sioux	Moon When Quilling and Beading Is Done
Haida	Kaganakyash Moon (no translation)
Kutenai	Falling River Moon
Laguna	Ripe Corn Moon
Lakota Sioux	Moon of the Changing Season
Micmac	Fat, Tame Animals Moon
Mohawk	Time of Poverty
Natchez	Turkeys Moon

Nez Perce	Falling Leaves Time
Ojibway	Falling of the Leaves Moon
Osage	Deer Breeding Moon
Oto	Deer Rutting Season Moon
Ponca	They Store Food in Caches Moon
San Ildefonso	Harvest Moon
San Juan	Leaf Fall Moon
Taos	Corn Ripe Moon
Tlingit	Big Moon
Wisham	Her Leaves Moon (also Travel in Canoes Moon)
Zuni	Great Sandstorm Moon (same as April)

NOVEMBER The second full moon after the fall equinox (in some but not all years).

Colonial American	Beaver Moon (also Frosty Moon, Snow Moon)
Algonquin	Beaver Moon
Cherokee	Month of the Trading Moon
Cheyenne	Freezing Moon
Choctaw	Sassafras Moon
Dakotah Sioux	Moon When Horns Are Broken Off, Moon When Deer Copulate
Haida	Stomach Moon
Kutenai	Killing Deer Moon
Laguna	Autumn Moon
Lakota Sioux	Moon of the Falling Leaves
Micmac	Tomcod Moon
Mohawk	Time of Much Poverty
Natchez	Bison Moon
Nez Perce	Autumn Time
Ojibway	The Freezing Up Moon
Osage	Coon Breeding Moon
Oto	Every Buck Loses His Horns Moon
Ponca	Beginning of Cold Weather Moon
San Ildefonso	All Gathered Moon
San Juan	All Gathered Moon

122

Taos	Corn Harvest Moon
Tlingit	Moon When People Have to Shovel Snow Away from Their Doors
Wisham	Her Frost Moon (also Snowy Mountains in the Morning Moon)
Zuni	Moon No Name Moon (same as May)

DECEMBER The full moon closest to the winter solstice.

Colonial American	Christmas Moon or Cold Moon (also Christ's Moon, Long Night Moon, Moon before Yule, Oak Moon)
Algonquin	Cold Moon (also Long Night's Moon)
Cherokee	Month of the Snow Moon
Cheyenne	Big Freezing Moon
Choctaw	Peach Moon
Dakotah Sioux	Twelfth Moon
Haida	Kungyadikadas (no translation)
Kutenai	Nistamu Natanik (no translation)
Laguna	Middle Winter Moon
Lakota Sioux	Moon of the Popping Trees
Micmac	The Chief Moon
Mohawk	Time of Cold
Natchez	Bears Moon
Nez Perce	Heekui (no translation)
Ojibway	The Little Spirit Moon
Osage	Baby Bear Moon
Oto	Cold Month Moon
Ponca	Beginning of Cold Weather with Snow Moon
San Ildefonso	Ashes Fire Moon
San Juan	Ashes Fire Moon
Taos	Night Moon
Tlingit	Ground-Hog Mother's Moon
Wisham	Her Winter Houses Moon
Zuni	Turning Moon (same as June)

THE HARVEST MOON.

SHINE ON, HARVEST MOON

In modern times, the full moon in September is widely referred to as the Harvest Moon, the only full moon that maintains such a traditional title. As a tradition passed down from European cultures, however, the Harvest Moon is actually the full moon closest to the fall equinox or the one which falls in the last week in September. But the full moon closest to this date can be either before of after the equinox, and can occur in October as well as September. The last October variation happened in 2006, as it will again in 2017. Between 1970 and 2050, the Harvest Moon will occur in October 18 times. The earliest that this prominent full moon occured in the same time span: September 8, in 1976 and the latest, on October 7, 1987.

ILLUSTRATION "The Harvest Moon," a print from a painting by Edward Duncan, c. 1864.

RARE MOONS

A blue moon is defined as a full moon that occurs twice in the same calendar month. When the date for one full moon falls on or near the beginning of a calendar month, the following full moon—always about 29½ days later—comes before the end of the month. February has only 29 days (except for leap years), so there can never a blue moon in this month. Blue moons occur approximately seven times every 19 years, an average of once every 33 months.

In the mid-1800s, atmospheric pollution from forest fires and volcanic eruptions was linked to a real event, when the Moon actually appeared to turn a shade of blue (it happened again in September 1950 in England) but this rare occurrence was not the origin of the term. The underlying origin of the phrase "once in a blue moon" dates to the Christian church in Europe in the Middle Ages, when church calendars traditionally included dates for the full moons—twelve in a "normal" year, but requiring an extra, or thirteenth full moon in one season (three months) every few years because of the differences between lunar cycles and the solar-based calendar. Each of the regular twelve full moons had a traditional name—Harvest Moon, Egg Moon, Wolf Moon, etc.—but when the thirteenth full moon occurred, it required a unique name of its own. Informally, this was the "Blue Moon," and because it was unusual and out of place, was considered unlucky, also lending an air of misfortune to the number thirteen. Why "blue"? In Middle English, the word *belew*—one of the predecessors of the modern word—meant "false" or "betrayer."

Calling a second full moon in a calendar month a blue moon did not arise until the mid-1900s. An error in *Sky & Telescope* magazine in March 1946, appears to be the trigger. An article in that issue attributed the term "blue moon" to a tradition used in the *Maine Farmers Almanac* to signify the occurrence of a fourth full moon in a season, which typically had only three (one season spans three months). In the *Maine Farmers Almanac* for 1937, the editors attributed the concept to the church calendars of the Middle Ages, as described above. However, in the *Sky & Telescope* article, the meaning for blue moon as used in calendars was mistakenly explained as an extra full moon that occurred in any single month—and the faulty usage stuck.

Blue moons have no significance in astronomy. Intriguingly, those months that do have blue moons are the same periods recognized as "leap months" in two older cultures. Traditional Chinese and Hindu calendars are based on a lunisolar system, and in order not to get out of step, these calendars

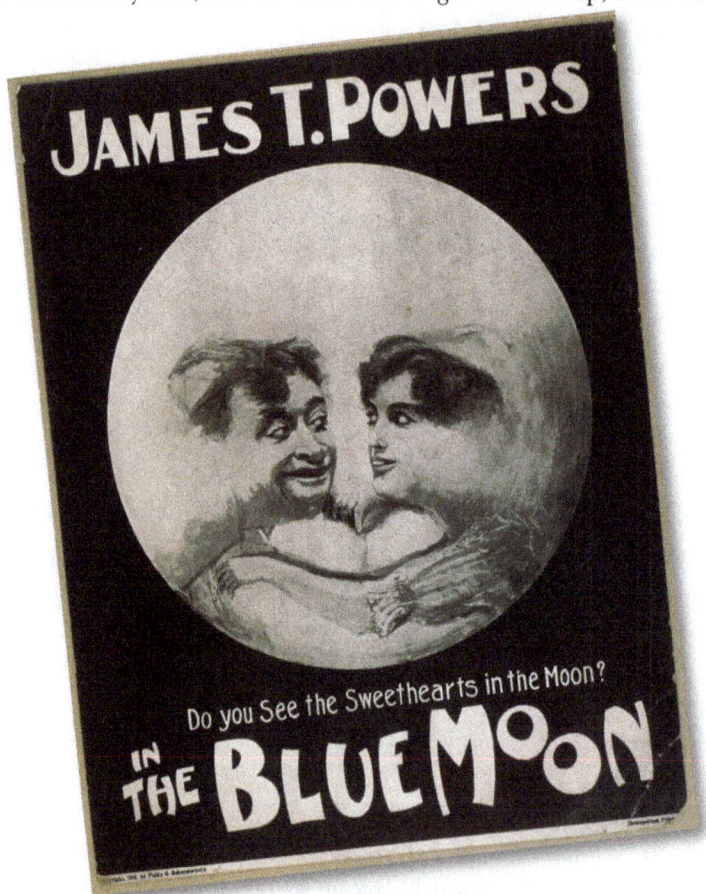

In 1906, *In the Blue Moon*, a musical comedy production was staged in a variety of U.S. cities, starring James T. Powers and Ethel Jackson. Here, the phrase uses the term "blue" to refer to a mood, not a second full moon in a month; historically, it was also used to mean "false" or "betrayal."

ILLUSTRATION "In the Blue Moon," a lithographed poster published in 1906. Courtesy of the Library of Congress.

must periodically adjust dates; the adjustment periods are the same months in which there are blue moons.

Blue moons may sometimes cause confusion because of differences in local time zones. A full moon that falls on the first day of the month but only a few hours after midnight, for example, will produce a second full moon in that month in the same time zone. But a few time zones to the west, the local time of the first full moon will actually fall on the last day of the preceding month, making the blue moon a month earlier.

Less noticeable than blue moons but equally rare are months where there are two new moons; the second is unofficially called a "black moon." In the twenty-first century, such a rare month will occur in 2018, 2037, 2067,

BLUE MOON VITAL STATISTICS

- One every 2.7 years
- 7 times every 19 years
- Once every 33 months
- Once every 33 full moons
- An average of 37 per century

BLUE MOON DATES

- January 31, 2018
- March 31, 2018
- October 31, 2020
- August 31, 2022
- May 31, 2026
- December 31, 2028
- September 30, 2031
- July 31, 2034

and 2094. Most of these events are also notable in that the month after will both have blue moons.

Even rarer is a month when there are no full moons. This can only happen in February and the event will only happen four times in the 1900s, the last was 1999. About the rarest of all blue moon events is a year with two. The last time this happened was in 1999 (January 31 and March 31) and the next time will be in 2018 (also in January and March). About as rare as a double blue moon is a full moon that falls on a leap day—February 29th—which happens about once every 100 years.

The last leap day full moon was in 1972; the next will occur in 2048, with subsequent dates in 2132, 2216, and 2376. As for leap years with no full moons in February, that last occurred in 1608; the next will be in 2572.

MOON OF MANY FACES

In some western cultures, popular legends and mythology describe a "man in the moon." This familiar image comes from the unique markings seen on the face of the moon, light and dark areas on the surface that seem to form the shape of a human face. In some cultures, different images characterize the surface patterns, including a "lady in the moon" and a "rabbit in the moon."

Other animals with claims to the lunar image include a beetle, a toad, a fox, a cow, a cat, a bear, and a lion. Sometimes, the patterns on the surface suggest different images at different stages in the Moon's path across the sky, because the arc of the path "tilts" the Moon relative to observers on Earth.

LADY IN THE MOON

MAN IN THE MOON

BEETLE IN THE MOON

RABBIT IN THE MOON

128

Noted photographer Edward S. Curtis took this picture of Kwakiutl dancers in 1914. The dance was held during a solar eclipse, an event when the tribe traditionally believed a sky creature had swallowed the Moon. The ritual was intended to cause the sky creature to sneeze, regurgitating the Moon.

Photograph courtesy of the Rare Book and Special Collections Division, Library of Congress.

VITAL STATISTICS OF THE MOON

MEAN DISTANCE OF THE MOON FROM EARTH	238,712 miles (384,400 km) 60.27 Earth radii 0.002 570 a.u.
GREATEST DISTANCE OF THE MOON FROM EARTH (APOGEE)	252,586 miles (406,740 km)
SHORTEST DISTANCE OF THE MOON FROM EARTH (PERIGEE)	221,331 miles (356,410 km)
CIRCUMFERENCE	6,790 miles (10,930 km) 0.27 of Earth's circumference
DIAMETER	2,160 miles (3,476 km) 0.27 of Earth's diameter
MEAN RADIUS	1,079 miles (1,737.5 km)
EQUATORIAL RADIUS	1,079 miles (1,738 km)
POLAR RADIUS	1,077 miles (1,735 km)
MEAN ANGULAR DIAMETER	31' 07"
MASS	8 x 1019 tons (7.35 x 1022 kg) 0.0123 Earth's mass
MASS RATIO (EARTH/MOON)	81.301
VOLUME	2.4 x 109 miles^3 (2.197 x 1010 km^3) 0.0203 Earth's volume
MEAN DENSITY	208 lb/ft^3 (3.34 g/cm^3) 3.33 more dense than water 0.6 Earth's density
GRAVITY AT THE SURFACE	5.31 ft/sec^2 (1.62 m/s^2) 0.1667 g (1/6 Earth's gravity)
ESCAPE VELOCITY	1.48 miles/sec (2.38 km/sec)
MEAN INCLINATION TO LUNAR EQUATOR	6° 41'

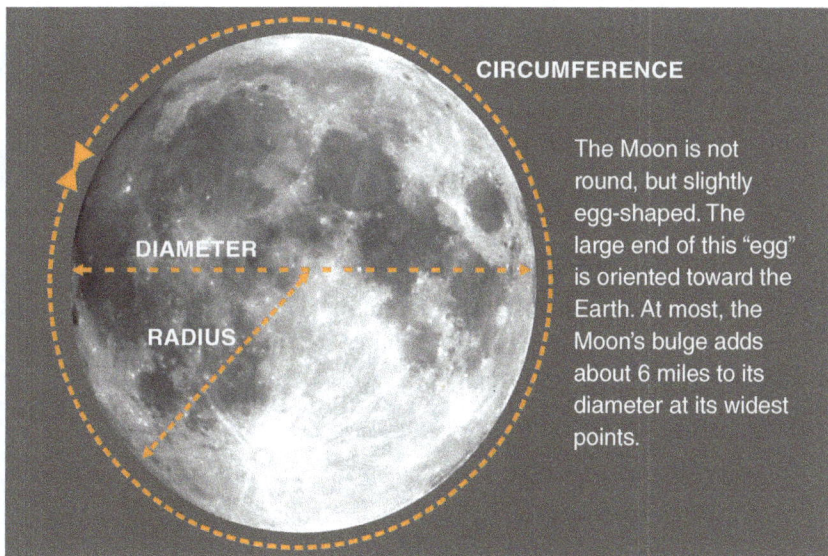

CIRCUMFERENCE

DIAMETER

RADIUS

The Moon is not round, but slightly egg-shaped. The large end of this "egg" is oriented toward the Earth. At most, the Moon's bulge adds about 6 miles to its diameter at its widest points.

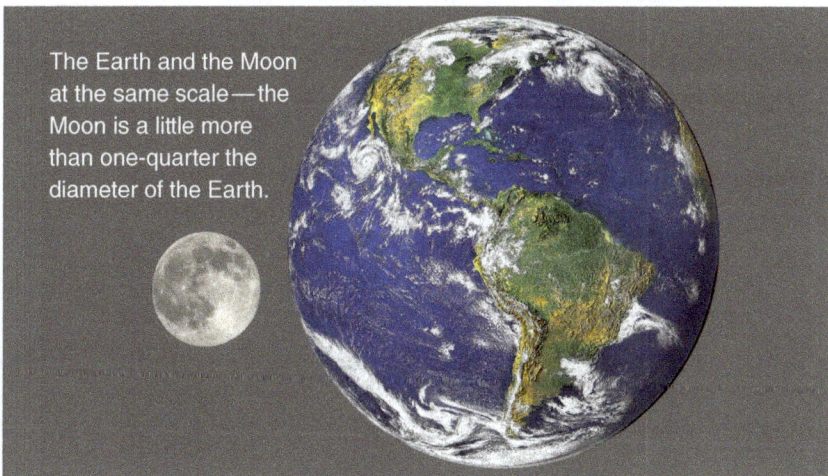

The Earth and the Moon at the same scale—the Moon is a little more than one-quarter the diameter of the Earth.

MEAN ORBITAL INCLINATION TO ECLIPTIC	5° 08' 43" (5.145°)
OSCILLATION OF ORBITAL INCLINATION TO EQUATOR	± 0° 9' every 173 days
INCLINATION OF LUNAR EQUATOR TO ECLIPTIC	1° 32' 33"
PERIOD OF REVOLUTION OF PERIGEE	3,232 days
ORBITAL DIRECTION	east (counterclockwise)
MEAN ORBITAL SPEED	2,287 miles/hour (3,683 km/hr) 33 minutes arc/hour
MAXIMUM ORBITAL VELOCITY	(1.076 km/sec)
MINIMUM ORBITAL VELOCITY	(0.964 km/sec)
DAILY SIDEREAL MOTION	13.176358 degrees
MEAN CENTRIPETAL ACCELERATION	0.01 ft/sec^2 (0.00272 m/sec^2) 0.0003 g
MEAN ECCENTRICITY OF ORBIT	0.0549 (mean eccentricity of the Earth's orbit is 0.0167)
SYNODIC MONTH (NEW MOON TO NEW MOON)	29.53059 days 29 days, 12 hr, 44 min, 2.8 sec 713 hours
SIDEREAL MONTH (STAR TO STAR)	27.32166 days 27 days, 7 hr, 43 min, 11.5 sec
ANOMALISTIC MONTH (APOGEE TO APOGEE OR PERIGEE TO PERIGEE)	27.55455 days 27 days, 13 hr, 18 min, 33.2 sec
NODICAL MONTH, DRACONIC MONTH (NODE TO NODE)	27.21222 days 27 days, 5 hr, 5 min, 35.8 sec
TROPICAL MONTH	27.321582 days

The diameter of the Moon or any other distant object can be measured in degrees of angle. This measurement is referred to as its angular diameter.

Depending on where it is in its elliptical orbit, the angular diameter of the Moon varies from 29'33" to 33'6", about one-half a degree. By comparison, at arm's length the human fist measures 10 degrees and a single finger represents about 2 degrees.

From the position of an observer, a vertical line passing through the center of the Earth generates two key points, a zenith and a nadir.

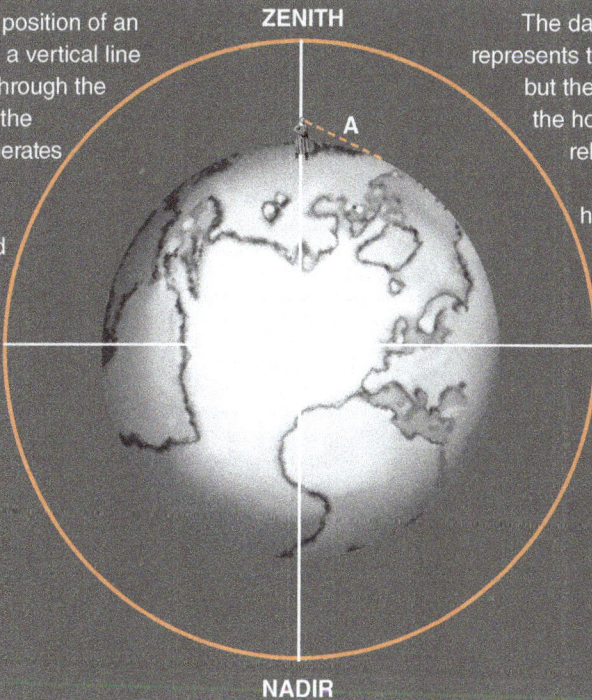

ZENITH

A

The dashed line A represents the horizon, but the position of the horizon shifts relative to the observer's height above the surface.

NADIR

(FIRST POINT OF ARIES TO FIRST POINT OF ARIES)	7 days, 7 hr, 43 min, 4.7 sec
REGRESSION OF NODES	6,798 days 18.6247 years (19.538 degrees per year)
ROTATION PERIOD	27.321661 days 27 days, 7 hr, 43 min, 11.5 sec
SURFACE TEMPERATURE	253° F (123° C) day –244° F (–153° C) night
SURFACE AREA	14,657,449 miles^2 (37,958,621 km^2) 9.4 billion acres 26% larger than Africa
VISIBLE SURFACE	50 percent during one lunar cycle 18 percent additional surface visible due to librations 59 percent total visible surface
PARALLAX	0.9507 degrees
MOON'S ANGULAR DIAMETER	0.5181 degrees
APPARENT MAGNITUDE OF FULL MOON	–12.5
AVERAGE ALBEDO	0.11
ESTIMATED AGE OF MOON	4.6 billion years
FLIGHT TIME FROM EARTH	60 to 70 hours
INCREASE IN MEAN DISTANCE FROM EARTH	1½ inches/year (3.8 cm/year)
TOTAL MASS OF ATMOSPHERE	~25,000 kg
SURFACE PRESSURE (NIGHT)	3 x 10^{-15} bar (2 x 10^{-12} torr)
ESTIMATED ATMOSPHERE [particles/cm^3]	Helium-4 (^{4}He) [40,000] Neon-20 (^{20}Ne) [40,000] Hydrogen (H^2) [35,000] Argon-40 (^{40}Ar) [30,000] Neon-22 (^{22}Ne) [5,000]

The Moon's orbital distance at apogee is the equivalent of 32 Earth diameters.

The Moon's orbital distance at perigee is the equivalent of 28 Earth diameters.

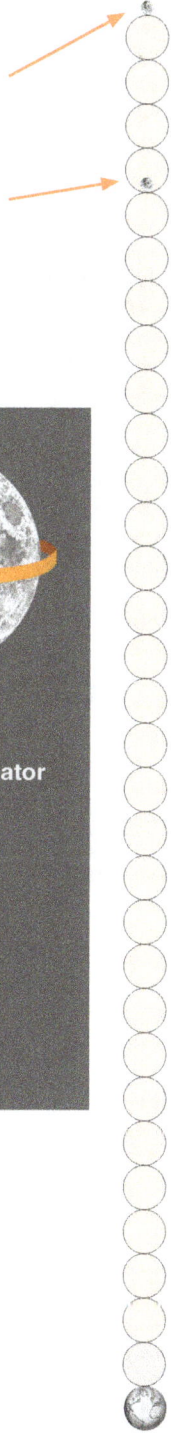

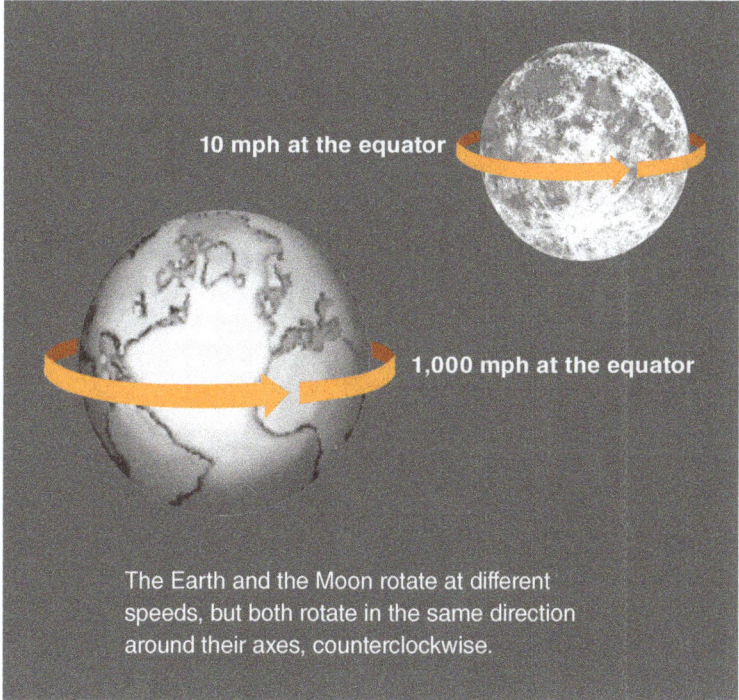

10 mph at the equator

1,000 mph at the equator

The Earth and the Moon rotate at different speeds, but both rotate in the same direction around their axes, counterclockwise.

Argon-36 (^{36}Ar) [2,000]
Methane   [1000]
Ammonia   [1000]
Carbon Dioxide (CO_2)   [1000]
Trace Oxygen (O^+)

AVERAGE CRUST THICKNESS	(68 km)
THINNEST CRUST	~0 m (under Mare Crisium)
THICKEST CRUST	(107 km) (on far side)
MAGNETIC FIELD	none

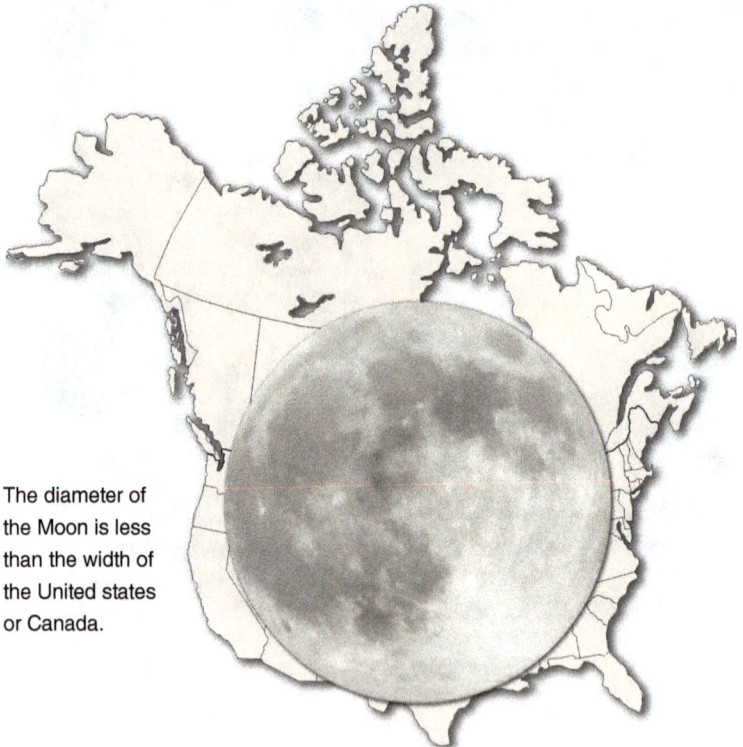

The diameter of the Moon is less than the width of the United states or Canada.

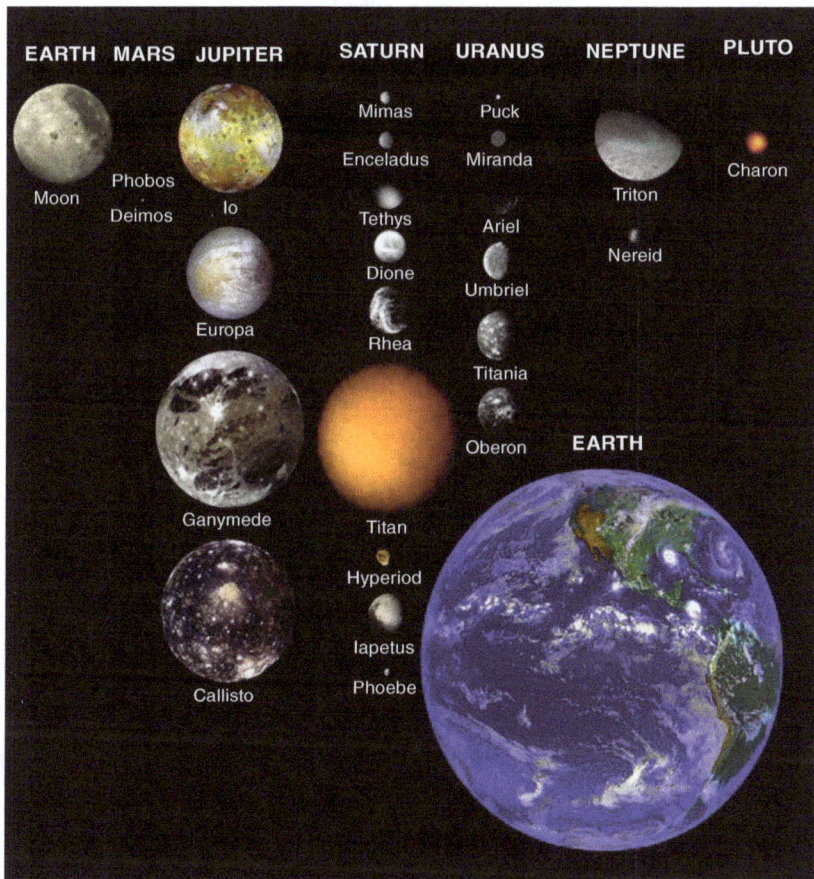

EARTH'S MOON IN SCALE TO THE MOONS OF THE SOLAR SYSTEM

Modified from an original NASA graphic.

WEIGHT ON THE MOON

EARTH WEIGHT	MOON WEIGHT	EARTH WEIGHT	MOON WEIGHT	EARTH WEIGHT	MOON WEIGHT
50 lbs	8.50 lbs	86	14.62	122	20.74
51	8.67	87	14.79	123	20.91
52	8.84	88	14.96	124	21.08
53	9.01	89	15.13	125	21.25
54	9.18	90	15.30	126	21.42
55	9.35	91	15.47	127	21.59
56	9.52	92	15.64	128	21.76
57	9.69	93	15.81	129	21.93
58	9.86	94	15.98	130	22.10
59	10.03	95	16.15	131	22.27
60	10.20	96	16.32	132	22.44
61	10.37	97	16.49	133	22.61
62	10.54	98	16.66	134	22.78
63	10.71	99	16.83	135	22.95
64	10.88	100	17.00	136	23.12
65	11.05	101	17.17	137	23.29
66	11.22	102	17.34	138	23.46
67	11.39	103	17.51	139	23.63
68	11.56	104	17.68	140	23.80
69	11.73	105	17.85	141	23.97
70	11.90	106	18.02	142	24.14
71	12.07	107	18.19	143	24.31
72	12.24	108	18.36	144	24.48
73	12.41	109	18.53	145	24.65
74	12.58	110	18.70	146	24.82
75	12.75	111	18.87	147	24.99
76	12.92	112	19.04	148	25.16
77	13.09	113	19.21	149	25.33
78	13.26	114	19.38	150	25.50
79	13.43	115	19.55	151	25.67
80	13.60	116	19.72	152	25.84
81	13.77	117	19.89	153	26.01
82	13.94	118	20.06	154	26.18
83	14.11	119	20.23	155	26.35
84	14.28	120	20.40	156	26.52
85	14.45	121	20.57	157	26.69

EARTH WEIGHT	MOON WEIGHT	EARTH WEIGHT	MOON WEIGHT	EARTH WEIGHT	MOON WEIGHT
158	26.86	195	32.15	55	9.35
159	27.03	196	33.32	56	9.52
160	27.20	197	33.49	57	9.69
161	27.37	198	33.66	58	9.86
162	27.54	199	33.83	59	10.03
163	27.71	200	34.00	60	10.20
164	27.88			61	10.37
165	28.05	25 kg	4.25 kg	62	10.54
166	28.22	26	4.42	63	10.71
167	28.39	27	4.59	64	10.88
168	28.56	28	4.76	65	11.05
169	28.73	29	4.93	66	11.22
170	28.90	30	5.10	67	11.39
171	29.07	31	5.27	68	11.56
172	29.24	32	5.44	69	11.73
173	29.41	33	5.61	70	11.90
174	29.58	34	5.78	71	12.07
175	29.75	35	5.95	72	12.24
176	29.92	36	6.12	73	12.41
177	30.09	37	6.29	74	12.58
178	30.26	38	6.46	75	12.75
179	30.43	39	6.63	76	12.92
180	30.60	40	6.80	77	13.09
181	30.77	41	6.97	78	13.26
182	30.94	42	7.14	79	13.43
183	31.11	43	7.31	80	13.60
184	31.28	44	7.48	81	13.77
185	31.45	45	7.65	82	13.94
186	31.62	46	7.82	83	14.11
187	31.79	47	7.99	84	14.28
188	31.96	48	8.16	85	14.45
189	32.13	49	8.33	86	14.62
190	32.30	50	8.50	87	14.79
191	32.47	51	8.67	88	14.96
192	32.64	52	8.84	89	15.13
193	32.81	53	9.01	90	15.30
194	32.98	54	9.18		

UT UNIVERSAL TIME	CIVIL TIME	EST EASTERN STANDARD TIME -5 hours	CST CENTRAL STANDARD TIME -6 hours	MST MOUNTAIN STANDARD TIME -7 hours	PST PACIFIC STANDARD TIME -8 hours
0:00	12:00 AM	7:00 PM	6:00 PM	5:00 PM	4:00 PM
1:00	1:00 AM	8:00 PM	7:00 PM	6:00 PM	5:00 PM
2:00	2:00 AM	9:00 PM	8:00 PM	7:00 PM	6:00 PM
3:00	3:00 AM	10:00 PM	9:00 PM	8:00 PM	7:00 PM
4:00	4:00 AM	11:00 PM	10:00 PM	9:00 PM	8:00 PM
5:00	5:00 AM	12:00 AM	11:00 PM	10:00 PM	9:00 PM
6:00	6:00 AM	1:00 AM	12:00 AM	11:00 PM	10:00 PM
7:00	7:00 AM	2:00 AM	1:00 AM	12:00 AM	11:00 PM
8:00	8:00 AM	3:00 AM	2:00 AM	1:00 AM	12:00 AM
9:00	9:00 AM	4:00 AM	3:00 AM	2:00 AM	1:00 AM
10:00	10:00 AM	5:00 AM	4:00 AM	3:00 AM	2:00 AM
11:00	11:00 AM	6:00 AM	5:00 AM	4:00 AM	3:00 AM
12:00	12:00 PM	7:00 AM	6:00 AM	5:00 AM	4:00 AM
13:00	1:00 PM	8:00 AM	7:00 AM	6:00 AM	5:00 AM
14:00	2:00 PM	9:00 AM	8:00 AM	7:00 AM	6:00 AM
15:00	3:00 PM	10:00 AM	9:00 AM	8:00 AM	7:00 AM
16:00	4:00 PM	11:00 AM	10:00 AM	9:00 AM	8:00 AM
17:00	5:00 PM	12:00 PM	11:00 AM	10:00 AM	9:00 AM
18:00	6:00 PM	1:00 PM	12:00 PM	11:00 AM	10:00 AM
19:00	7:00 PM	2:00 PM	1:00 PM	12:00 PM	11:00 AM
20:00	8:00 PM	3:00 PM	2:00 PM	1:00 PM	12:00 PM
21:00	9:00 PM	4:00 PM	3:00 PM	2:00 PM	1:00 PM
22:00	10:00 PM	5:00 PM	4:00 PM	3:00 PM	2:00 PM
23:00	11:00 PM	6:00 PM	5:00 PM	4:00 PM	3:00 PM

TIME CONVERSIONS   PREVIOUS DAY

RESOURCES

CALENDARS AND ANNUALS

The Astronomical Companion. Created by Guy Ottewell. A large format publication with extensive illustrations, diagrams, and graphs depicting a wide range of celestial phenomena, including much useful perspective on the Moon. Supported and available through the author's Web site, and select retail outlets.
www.universalworkshop.com/ACOM.htm

Astronomical Phenomena. Usually available 1–2 years in advance of a calendar year. The major resource for listings of specific solar and lunar cycles, including phases, settings, and risings. Published by the U.S. Naval Observatory; most contents directly accessible through the USNO Web site. For sale from the U.S. Government Printing Office. www.usno.navy.mil

The Moon Calendar. Created and produced annually by the author of *The Moon Book.* Issued annually beginning in August preceding the calendar year. Card-size display of the individual phases of the Moon for every day of the year. Available at local bookstores, museum stores, and telescope stores. The Experiment Publishing
www.theexperimentpublishing.com

Observer's Handbook. Issued annually in the fall preceding the calendar year. A detailed collection of astronomical events, including moon cycles, rising, setting, occultations, librations, and eclipses. Limited distribution in the U.S. Royal Astronomical Society of Canada. www.rasc.ca

The Old Farmer's Almanac. Issued annually (usually in September preceding the calendar year) by Yankee Publishing. Includes moon rise and moon set times as well as other basic astronomical data. For sale at most newsstands and bookstores. Yankee Publishing. www.almanac.com

PERIODICALS

Astronomy Magazine. Kalmbach Publishing Company www.astronomy.com

Griffith Observer. Griffith Observatory www.griffishobservatory.org

The Planetarian. International Planetarian Society www.ips-planetarium.org

Sky and Telescope Magazine. Sky Publishing Corporation www.skyandtelescope.com

Sky News: The Canadian Magazine of Astronomy & Stargazing. National Museum of Science and Technology www.skynews.ca

The Strolling Astronomer. Official publication of the Association of Lunar and Planetary Observers www.alpo-astronomy.org/djalpo

ORGANIZATIONS

American Astronomical Society www.aas.org

Association of Lunar and Planetary Observers www.alpo-astronomy.org

Astronomical League wwww.astroleague.org

Astronomical Society of the Pacific www.astrosociety.org

Lunar & Planetary Institute www.lpi.usra.edu

National Space Society www.nss.org

Royal Astronomical Society www.ras.org.uk

Royal Astronomical Society of Canada www.rasc.ca

MAPS / GLOBES

Earth's Moon Wall Map. This large color map has labeled details of both front and back sides of the Moon. Published by the National Geographic Society shop.
nationalgeographic.com

NASA Lunar Chart (LPC-1). The Defense Mapping Agency Aerospace center created this map for NASA. In a large color format, the map includes a detailed mercator projection of the surface between 45° north and 45° south, as well as stereographic projections of both polar regions. Downloadable jpg.
www.hq.nasa.gov/alsj/LPC-1.html

Field Map of the Moon. The publisher of Sky & Telescope Magazine has also created this small (12 inch diameter) map of major features of the lunar surface. Published by Sky Publishing Corporation
www.shopatsky.com/sky-and-telescope-field-map-of-the-moon

Moon Globe. 12" diameter globe with lunar features. Published by Rand McNally store. www.randmcnally.com
Sky & Telescope's Moon Globe. 12" diameter globe with lunar features.
www.shopatsky.com/moon-globe

Government Maps. The U.S. Geological Survey is the official agency responsible for maps of the lunar surface, as well as other printed maps of celestial bodies.
www.astrogeology.usgs.gov/maps

SOFTWARE / APPS

Software that has been created about astronomy, the stars, or the solar system often includes information, images, and interactive features about the Moon and its cycles, typically permitting users to adjust settings for locations and dates. Planetarium programs also frequently include such interactive features. New programs may be discovered from reviews and announcements in astronomy publications such as Sky & Telescope and Astronomy, as well as online astronomy Web sites and blogs. Some services, software, and apps are free or shareware.

Astronomy Lab 2 (Win). www.ericbt.com/astronomylab2

Distant Suns (Win/Mac app). www.distantsuns.com

Deep-Sky Planner (Win). knightware.biz/dsp

LunarPhase Pro (Win) www.lunarphasepro.nightskyobserver.com

MoonMenu (Win/Mac) www.selznick.com/products/moonmenu

Moonphase (Win/Mac) www.moonphase.en.softonic.com

Moon rise (Win). www.moonrise.us/moonrise.html

Multiyear Interactive Computer Almanac (Win/Mac). http://aa.usno.navy.mil/software/mica/micainfo.php

QuickPhase Pro Moon Phase (Win/Mac) www.moonconnection.com/quickphase

The Sky (Win/Mac) www.bisque.com

Stellarium (Win/Mac) www.stellarium.en.softonic.com

Sun & Moon Calculator (Win/Mac) sun-moon-calculator.software.informer.com

Time & Date Moon Phases (Win/Mac) www.timeanddate.com/calendar/moonphases.html

Voyager (Win/Mac). www.carinasoft.com/voyager.html

ONLINE

The contents and organization of online sites provide significant information about the Moon, including history, geology, exploration, photography, and lunar cycles. Some of the sites listed here may change names or move to new Internet addresses, and new resources can become available at any time. For the additional results, search with one or more appropriate keywords: moon, lunar, eclipse, moon rise, phase, etc.

Adler Planetarium astro.uchicago.edu/adler

American Astronomical Society www.aas.org

Association of Lunar & Planetary Observers www.jpl.arizona.edu/alpo

Astronomical League www.mcs.net/~bstevens/al

Astronomical Society of the Pacific www.aspsky.org

Astronomy Magazine www.kalmbach.com/astro/astronomy.htm

AstroPixels www.astropixels.com

EarthSky www.earthsky.org/moon-phases

Farmers' Almanac www.farmersalmanac.com

Griffith Observatory www.GriffithObs.org

Guy's Blog universalworkshop.com/guysblog

International Occultation Timing Assoc. www.lunar-occultations.com

In-The-Sky www.in-the-sky.org

Jet Propulsion Laboratory www.jpl.nasa.gov

Lunar & Planetary Society cass.jsc.nasa.gov

The Lunar Ephemeris twitter.com/LunarDaily

Lunar Phases Watch twitter.com/lunatic_watch

The Moon Almanac www.themoonalmanac.wordpress.com

NASA www.nasa.gov

National Space Science Center nssdc.gsfc.nasa.gov

North American Skies twitter.com/NASkies

The Old Farmer's Almanac www.almanac.com

Sky & Telescope Magazine www.skypub.com

Solar & Lunar Eclipses planets.gsfc.nasa.gov/eclipseeclipse.html

Time and Date www.timeanddate.com/moon

U.S. Naval Observatory www.usno.navy.mil

TELESCOPES / ASTRONOMY GEAR

Barska www.barska.com

Bresser bresserusa.com

Bushnell www.bushnell.com

Carson Optical www.carson.com/telescopes

Celestron Telescopes www.celestron.com

Edmund Scientific Company www.scienfiticsonline.com

Explore Scientific USA www.explorescientificusa.com

iOptron www.ioptron.com

Konus Optical & Sports systems www.konus.com

Levenhuk www.levenhuk.com

Meade Instruments www.meade.com

Olivon www.olivonmanufacturing.com

Orion Telescopes www.telescope.com

Santa Barbara Instrument Group (SBIG) www.sbig.com

Sky-Watcher www.skywatcher.com

Software Bisque www.bisque.com

Tele Vue Optics www.televue.com

Vixen Optics www.vixenoptics.com

Zeiss www.zeiss.com

Zhumell www.xhumell.com

BIBLIOGRAPHY

Alter, Dinsmore. *Pictorial Guide to the Moon*, Third Revised Edition. 1973, Thomas Y. Crowell Company (New York).

Bowditch, Nathaniel. *American Practical Navigator*. 1977, Defense Mapping Agency Hydrographic Center.

Bredon, Juliet, and Mitrophanow, Igor. *The Moon Year: A Record of Chinese Customs and Festivals*. 1927, Kelly & Walsh, Ltd. (Shanghai).

British Astronomical Association. *Guide to Observing the Moon*. 1986, Enslow Publishers, Inc. (Hillside, NJ).

Brunner, Bernd. *Moon: A Brief History*. 2010, Yale University Press.

Cadogan, Peter H. *The Moon: Our Sister Planet*. 1981, Cambridge University Press.

Carr, Michael H., editor. *The Geology of the Terrestrial Planets*. 1984, National Aeronautics and Space Administration.

Corliss, William R. *The Moon and the Planets: A Catalog of Astronomical Anomalies*. 1985, The Sourcebook Project (Glen Arm, Maryland).

Davidson, Norman. *Sky Phenomena: A Guide to Naked-Eye Observation of the Stars*. 1993, Lindisfarne Press (Hudson, NY).

Dershowitz, Nachum, and Reingold, Edward M. *Calendrical Calculations*, Third Edition. 2008, Cambridge University Press.

Duffet-Smith, Peter. *Practical Astronomy with Your Calculator, Second Edition*. 1981, Cambridge University Press.

Dunlop, Storm. *Astronomy, A Step-by-Step Guide to the Night Sky*. 1985, Macmillan Publishing Company.

Firsoff, V.A. *Strange World of the Moon*. 1959, Basic Books.

French, Bevan M. *The Moon Book*. 1977, Penguin Books.

Gater, Will, and Sparrow, Giles. *The Night Sky Month by Month*. 2011, DK Publishing.

Harrington, Philip S. *Eclipse!: The What, Where, When, Why & How Guide to Watching Solar & Lunar Eclipses*. 1997, John Wiley & Sons, Inc.

Heiken, Grant, Vaniman, David, and French, Bevan M. *Lunar Sourcebook: A User's Guide to the Moon*. 1991, Cambridge University Press.

Hockey, Thomas. *How We See the Sky: A Naked-Eye Tour of Day and Night*. 2011, University of Chicago Press.

Kaler, James B. *The Ever-Changing Sky: A Guide to the Celestial Sphere.* 1996, Cambridge University Press.

Kanipe, Jeff. *A Skywatcher's Year.* 1999, Cambridge University Press.

Katzeff, Paul. *Full Moons: Fact and Fantasy about Lunar Influence.* 1981, Citadel Press/Lyle Stuart Inc.

Kelly, Adrian, Dresser, Peter, and Ross, Linda M. *Religious Holidays and Calendars.* 1993, Omnigraphics (Detroit, MI).

Kelly, Patrick, editor. *Observer's Handbook,* various annual editions. The Royal Astronomical Society of Canada.

Martinez, Patrick, editor. *The Observer's Guide to Astronomy.* 1994, Cambridge University Press.

McCluskey, Stephen C. *Astronomies and Cultures in Early Medieval Europe.* 1998, Cambridge University Press.

Meeus, Jean. *Astronomical Algorhythms.* 1991, Willman-Bell, Inc.

Meeus, Jean. *Astronomical Formulae for Calculators,* Fourth Edition. 1988, Willman-Bell, Inc. (Richmond, VA).

Meeus, Jean. *Astronomical Tables of the Sun, Moon, and Planets,* Second Edition. 1995, Willman-Bell, Inc. (Richmond, VA).

Menzel, Donald H. *A Field Guide to the Stars and Planets.* 1964, Peterson Field Guide Series/Houghton Mifflin Company.

Moore, Patrick. *The Moon.* 1981, Mitchell Beazley Publishers/Rand McNally & Company.

Meeus, Jean. *Astronomical Tables of the Sun, Moon and Planets,* Second Edition. 1995, Willmann-Bell, Inc.

Muirden, James, editor. *Sky Watcher's Handbook.* 1993, W.H. Freeman and Company Limited.

Naylor, Ernest. Moonstruck: How Lunar Cycles Affect Life. 2015, Oxford University Press.

Ottewell, Guy. *The Astronomical Companion.* 1979, Universal Workshop at Furman University (Greenville, SC).

Parise, Frank. *The Book of Calendars.* 1982, Facts On File, Inc.

Price, Fred W. *The Moon Observer's Handbook.* 1988, Cambridge University Press.

Rackham, Thomas. *Moon in Focus.* 1968, Pergamon Press Ltd. (London, U.K.).

Reingold, Edward M. and Dershowitz, Nachum. *Calendrical Tabulations 1900–2200.* 2002, Cambridge University Press.

Richards, E.G. *Mapping Time: The Calendar and its History.* 1998, Oxford University Press.

Rükl, Antonín. *Atlas of the Moon.* 1990, Kalmbach Books (Waukesha, WI).

Seidelmann, P. Kenneth, editor. *Explanatory Supplement to The Astronomical Almanac.* 1992, University Science Books (Mill Valley, CA).

Sheehan, William P. and Dobbins, Thomas A. *Epic Moon: A History of Lunar Exploration in the Age of the Telescope.* 2001, Willman-Bell.

Spudis, Paul D. *The Once and Future Moon.* 1996, Smithsonian Institution Press.

Steel, Duncan. Marking Time: *The Epic Quest to Invent the Perfect Calendar.* 2000, John Wiley & Sons.

Turnill, Reginald. *The Moonlandings: An Eyewitness Account.* 2003, Cambridge University Press.

U.S. Naval Observatory. *Explanatory Supplement to the Astronomical Ephemeris and the American Ephemeris and Nautical Almanac.* 1961, U.S. Naval Observatory.

Westfall, John E. *Atlas of the Lunar Terminator.* 2000, Cambridge University Press.

Westrheim, Margo. *Calendars of the World: A Look at Calendars and the Ways We Celebrate.* 1993, Oneworld Publications (Oxford, England).

Whipple, Fred L. *Earth, Moon, and Planets,* Third Edition. 1970, Harvard University Press.

Whitaker, Ewen A. *Mapping and Naming the Moon.* 1999, Cambrige University Press.

Wood, Charles A. *The Modern Moon.* 2003, Sky Publishing.

Wylie, Francis E. *Tides and the Pull of the Moon.* 1979, Berkley Publishing Corp.

GLOSSARY

albedo
: The ratio of light reflected from the surface of a planet or Moon. Albedo is determined by measuring the ratio between the light reflected and the light shining on an object; complete reflection is represented by an albedo of 1.

altitude
: The distance in degrees of an object measured from the horizon up. Not the same unit as declination, which measures this distance from the celestial equator.

angular diameter
: The diameter of a distant object as measured by the angle formed from a point representing an observer and the outer edges of the object.

annular eclipse
: An eclipse of the Sun when the Moon is farthest away in its orbit around Earth. At this point, its apparent diameter is not large enough to completely obscure the sun. During an annular eclipse, a ring of light is left uncovered around the dark circle produced by the Moon.

anomalistic month
: The period of time it takes the Moon to go from one point of apogee (or perigee) to the next: 27.55455 days.

anorthositic rock
: One of the types of rocks found on the Moon at higher elevations.

aphelion
: The point in a planet's orbit around the Sun when it is farthest from the Sun (opposite of perihelion).

apogee
: The point in the Moon's orbit when it is farthest from the Earth (opposite of perigee).

apogean tide
: The low tide of the month that occurs when the Moon is at apogee (farthest from Earth).

apolune
: The point in the orbit of an object around the Moon (such as a spacecraft) when it is farthest from the Moon's surface.

asteroid
: A body of rock or frozen liquid that is in orbit around the Sun. Asteroids are sometimes considered planetoids.

149

astronomical unit (A.U.)	The mean distance between the Sun and the Earth and used as a standard of measurement. 1 A.U. = 92,955,630 miles (149,597,870 km).
astronomy	The science dealing with objects in space.
astrophysics	A branch of astronomy, using physics to study and explain celestial objects.
axis	An imaginary line through the center of mass of an object, around which the object rotates.
azimuth	The angle along the horizon measured from due north of an observer to directly under an object of interest in the sky. With the observer facing south, north is 0 degrees, east is 90 degrees, south is 180 degrees, and west is 270 degrees.
barycenter	A point marking the center of mass created when two celestial objects orbit around each other.
basalt	A type of igneous rock created from lava and found on the lunar surface in low areas, also found on Earth.
breccia	A composite rock found on the Moon and formed from small pieces of different minerals, also found on Earth.
bright limb	The illuminated outer edge of the Moon or planet.
CCD	Charge-coupled device. A camera or the component of a camera that registers and records images.
celestial equator	An imaginary extension of the Earth's equator into the sky. The celestial equator is 90 degrees from each of Earth's celestial poles.
celestial mechanics	The branch of astronomy dealing with the motions and gravitational effects of celestial objects.
circumference	The linear measurement around the outside of a circle or a sphere.

colongitude	The longitude on the surface of the Moon marked by the terminator, the edge of the area illuminated by the Sun.
conjunction	The position of two celestial bodies when they are in line with one another as seen by an observer on Earth. The new moon is also referred to as moon in conjunction with the Sun (opposite of opposition).
crater wall	The wall formed by the impact of a meteorite on the lunar surface, usually circular, but can be oblong or incomplete.
crescent moon	A phase of the Moon just before and after the new moon, when only a thin curved section is lighted by the Sun. The last crescent moon before the new moon is sometimes called the old crescent moon, and the first crescent moon after the new moon is sometimes called the young crescent moon.
culmination	The highest point a celestial body reaches in the sky as seen from a given location on Earth, always occurring when the body's azimuth is 180, or due south.
cusp	The extreme end of a bright limb.
dark limb	The dark, or unlit edge of the Moon or planet.
dark of the moon	Another name for the new moon.
Daylight Saving Time (DST)	A legislated time change in some countries in which local times are moved up by one hour in the spring and back one hour in the fall ("spring ahead, fall back").
declination	The angle measured between the celestial equator and an object in the sky, with the equator being 0° and the highest point at 90°.
density	An object's mass divided by its volume (grams/centimeter).
diurnal	Referring to a period of one day.

DST   See Daylight Saving Time.

earthshine   Reflected light from the Earth, visible as a dull, red, or copper glow on the Moon during lunar eclipses. Earth-shine can also sometimes illuminate a young crescent moon so that the whole face of the Moon can be faintly seen. This effect is often referred to as the "old moon in the new moon's arms."

eclipse   The blocking of light from the Sun when the Earth comes between the Sun and the Moon or the Moon between the Sun and the Earth.

ecliptic   The imaginary line formed by the Earth's orbit around the Sun or the plane formed by the apparent motion of the Sun through the sky.

elliptical orbit   A non-circular path formed when a body moves around another. The shape is that of a "stretched" or distorted circle.

elongation   The angle of a planet away from the Sun or the Moon from the Earth as viewed from the Earth.

ephemeris   A publication or list that has information needed to locate a star, moon, or planet in the sky at a particular time.

equatorial tide   A tide produced semi-monthly by the position of the Moon over the equator.

escape velocity   The speed required for an object to overcome the gravitational force of an astronomical object.

far side   The side of the Moon facing away from the Earth.

first quarter moon   The phase of the Moon when it is 90 degrees away from a line between the Sun and the Earth, measured eastward from the Sun, as seen from the north. The angle of illumination creates a half-circle picture of the Moon's surface, with the lighted half being on the right side.

full moon	The phase of the Moon when it is on the opposite side of the Earth from the Sun and receives sunlight across its entire face, forming a circle of light. At this point, the Moon is in opposition to the Sun.
gibbous moon	The phase of the Moon when it is getting larger after the first quarter moon phase (waxing gibbous) or smaller after the full moon but before the last quarter moon (waning gibbous).
gravity	One of the fundamental forces of nature, defined as the constant force of attraction between all objects in the universe. The gravitational force is inversely proportional to the square of the distance between the objects and proportional to the masses.
grazing occultation	An occultation by the Moon of a planet or star where the path of the planet or star only intercepts the north or south limb of the Moon.
Greenwich Mean Time	Time as measured from the 0 degrees longitude position of the Greenwich Observatory in England, also known as Universal Time (UT).
half moon	See first quarter moon or last quarter moon.
intercalation	A method of synchronizing a lunar calendar with a solar year by adding extra days or months. Extra days are known as intercalary days and extra months are known as intercalary months.
lacus	Latin for lake. An area on the surface of the Moon resembling a lake.
last quarter moon	The phase of the Moon when it is 90 degrees away from a line between the Sun and the Earth, measured westward from the Sun, as seen from the north. The angle of illumination creates a half circle of the Moon's surface, with the lighted half being on the left side. Also referred to as the third quarter moon.

latitude	Lines of measurement around a planet or the Moon, parallel to its equator. Measured in degrees, with the equator being 0 degrees and the poles 90 degrees north or south.
librations	The irregular motions of the Moon in its elliptical orbit around Earth that allow slightly more than half of the Moon's surface to be visible over a period of time.
limb	The illuminated, visible edge of a planet or moon.
longitude	Lines of measurement at right angles to the equator of a planet or the Moon. Measured in degrees of angle from a designated line of 0 degrees. On the Moon, 0 degrees longitude is at the center of the visible face, in the Sinus Medii.
lunar day	The period of time between two successive transits of the Moon over the same meridian. The mean lunar day is 24.84 hours (1.035 times the mean solar day). Not the same thing as a day on the Moon, which corresponds to a synodic month.
lunar eclipse	An eclipse created by the Earth coming between the Sun and the Moon. Lunar eclipses always happen during the full moon phase.
lunar interval	The elapsed time between the transit of the Moon over the Greenwich meridian and a local meridian.
lunar rays	Visible streaks on the surface of the Moon which radiate away from some craters.
lunartidal interval	The length of time between the transit of the Moon and following high or low tide.
magnitude	A numerical value indicating the brightness of an object in space; the higher the number, the dimmer the object.
mare	Latin for sea (plural: maria). An area on the surface of the Moon (or Mars) that is low, dark, and formed from ancient lava flows.

mascon An area of the Moon's surface formed from dense, thick lunar material and having strong local gravitational effects.

mean A mathematical average of a set of numbers or measurements, with the mean equaling the sum of the numbers divided by the number of units. The mean radius of the Moon, for example, is the average radius figured from multiple measurements.

meridian An imaginary line that passes directly north and south through an observer or specified location on Earth. A plane extended from this line into space passes through the zenith (point above the observer).

meteoroid A small body drifting through space. If a meteoroid is pulled into a planet's or moon's gravitational field and survives a trip through the atmosphere and lands, it is called a meteorite. If meteorites are not burned up in the process of entering an atmosphere (or if there is no atmosphere), they may strike the surface and if large enough, create craters. Technically, a "meteor" is just the visual streak of light made by a meteoroid as it passes through the atmosphere.

Moon The natural satellite of the Earth. Other planets have natural satellites as well, which may be referred to as "moons" without using a capital M.

moon rise The point in time when the upper limb of the Moon is even with the Earth's horizon as the Moon rises in the east.

moon set The point in time when the upper limb of the Moon is even with the Earth's horizon as the Moon sets in the west.

nadir An imaginary point directly under an observer on the surface of the Earth, extending through the Earth and into the sky (opposite of zenith).

neap tide The lowest high tide of the lunar month, occurring near the first and last quarter moon phases.

near side The side of the Moon facing the Earth.

new moon	The phase of the Moon when it is directly between the Earth and the Sun. Because sunlight is hitting only the far side of the Moon, it appears dark from the Earth. When the Moon is visible just before or after the new moon, the phase is called a crescent moon.
nodes	The imaginary points at which the orbital path of the Moon or other celestial body crosses the ecliptic.
nodical month	A lunar cycle measured by the Moon moving from one and back again: a period of 27.21222 days.
occultation	The movement of one celestial object behind another, such as the occultation of the star Spica by the Moon.
old crescent moon	Another name for the thin crescent of the Moon that is still illuminated by the Sun before the Moon goes completely dark at the new moon phase.
one-quarter moon	See first quarter moon.
opposition	A specific point in time when a moon or planet is 180 degrees away from the Sun, on the other side of the Earth. The Moon is full when it is in opposition (opposite of conjunction).
orbital eccentricity	The degree to which an elliptical orbit is elongated. Measured by the distance between the foci divided by the major axis.
palus	Latin for swamp. An area on the surface of the Moon that is dark.
parrallax	The perceived displacement of a distant object such as a moon, planet, or star due to the movement of the Earth.
partial eclipse	A lunar eclipse in which the Moon only partly enters the dark, umbral shadow of the Earth but is inside the secondary, penumbral shadow. Also refers to a solar eclipse when the Moon does not line up completely between the

Earth and Sun and only partly obscures the Sun. This type of eclipse also produces a penumbra as well as an umbra.

penumbra The lighter part of a shadow that is formed by diffused light in an area around the edges of an object.

perigee The point in the Moon's orbit when it is closest to the Earth (opposite of apogee).

perigean tide The high tide of the month that occurs when the Moon is at perigee (closest to to Earth).

perihelion The point in a planet's orbit around the Sun when it is closest to the Sun (opposite of aphelion).

perilune The point in the orbit of an object (such as a spacecraft) around the Moon when it is closest to the Moon's surface.

phases The visible changes that the Moon goes through in every lunar month, caused by the changing angle of illumination from the Sun. There are four specific phases—new moon, first quarter moon, full moon, and last quarter moon—and also non-specific phase names such as waxing moon, waning moon, gibbous moon, and crescent moon.

quadrature The position of the Moon or a planet when it is at right angles to the Sun. The Moon is in first quarter phase when it is in east quadrature to the Sun and last quarter phase when it is in west quadrature.

quarter moon The phase of the Moon that can be either the first quarter moon or the last quarter moon. This phase occurs when the Moon is 90 degrees away from a line between the Sun and the Earth. In the northern hemisphere, the angle of illumination creates a half circle picture of the Moon's surface, with the lighted half being on the right side during first quarter moon and on the left side for last quarter moon.

radius The linear measurement from the center of a sphere to the surface, or half of the diameter.

regression of nodes	The backwards movement of the Moon's nodes relative to direction of orbit.
revolution	The movement of one body around another in an orbit. Not to be confused with rotation.
right ascension	The measurement in a horizontal direction around the celestial equator, with the perpendicular axis passing through both the north and south celestial poles. Indicated in units of hours, minutes, and seconds with the zero point at the vernal equinox (the complete circle measures 360°, and one hour of right ascension is equal to 15°). Not to be confused with azimuth, which measures horizontal points from the plane of an observer.
rille	A valley or small canyon on the surface of the Moon.
rotation	The spinning of a body around its own axis. Not to be confused with revolution.
Saros Cycle	A cycle of lunar months lasting 18 years and 11.3 days, the time it takes the Moon, the Earth, and the Sun to return to the same position relative to each other.
satellite	An object that is in orbit around another object in space.
selenography	The science dealing with the study of the surface of the Moon.
selenology	The science dealing with the study of the Moon. From the Greek goddess, Selene.
sidereal month	A lunar month measured by a return to a specific position marked by a certain star: a period of 27.32166 days.
sinus	Latin for bay. An area on the surface of the Moon resembling the bay of an ocean.
solar eclipse	An eclipse caused when the Moon comes directly between the Earth and the Sun, temporarily blocking out Sun's disk in the sky.

spring tide	The highest tides in a lunar month, occurring near new and full moons, when the Earth, Sun, and Moon are aligned.
sunrise terminator	The line separating the light and dark segments of the Moon when the Moon is waxing. Also called the morning terminator.
sunset terminator	The line separating the light and dark segments of the Moon when the Moon is waning. Also called the evening terminator.
synodic month	A lunar month as measured from the point of one new moon to the next new moon: a period of 29.53059 days.
syzygy	In astronomy, a configuration of three or more celestial bodies in a straight line. Both lunar and solar eclipses—involving the Sun, the Earth, and the Moon—are syzygies. The term is also sometimes used even if there is not a perfectly straight line, as during a typical full or new moon that does not involve an eclipse.
tektites	Small particles made of glasslike material, formed from the impact of meteorites.
terminator	The line formed by the edge of the illuminated portion of the Moon.
three-quarter moon	See last quarter moon.
tide	The cyclical movement of bodies of water or land on the Earth or the Moon caused by the gravitational pull of the Earth, Moon, and Sun.
transit	The point when the path of the Moon, the Sun, a star, or a planet takes it across the meridian.
tropical month	The time required for the Moon to move from the first point of Aries and back again, a period of 27.321582 days.
umbra	The darker core of a shadow, usually cone shaped, and

surrounded by a lighter penumbral shadow. Also refers to the darker center of sunspots.

waning moon
The period in the Moon's monthly cycle after the full moon and before the new moon. During this period, the lighted portion of the Moon's surface is decreasing.

waxing moon
The period in the Moon's monthly cycle after the new moon and before the full moon. During this period, the lighted portion of the Moon's surface is increasing.

young crescent moon
Another name for the thin crescent of the Moon that is illuminated by the Sun just after the new moon.

zenith
The imaginary point directly above an observer on Earth (opposite of nadir).

ACKNOWLEDGMENTS

Cathie and Tim Havens (S&S Optika and Omega Haven)
Larry Sessions
The U.S. Naval Observatory
The Denver Public Library
Auraria Library, Metropolitan State College
Norlin Library, University of Colorado
The Bloomsbury Review (Denver, Colorado)
The Tattered Cover Bookstore (Denver, Colorado)
Gregory McNamee
Colleen Gino
Kathleen Cain
Special thanks in absentia to Leroy Doggett of the U.S. Naval
Observatory for his initial help in the conception and research
of *The Moon Calendar* and the first edition of *The Moon Book*.

INDEX

Earth rising above the Moon's horizon, taken during the Apollo 8 mission (the black and white original photo was recreated in color by NASA).